Springer Series in Geomechanics and Geoengineering

Editors: Wei Wu · Ronaldo I. Borja

Magued Iskander

Behavior of Pipe Piles in Sand

Plugging and Pore-Water Pressure Generation During Installation and Loading

Springer

Professor Wei Wu, Institut für Geotechnik, Universität für Bodenkultur, Feistmantelstraße 4, 1180 Vienna, Austria, E-mail: wei.wu@boku.ac.at

Professor Ronaldo I. Borja, Department of Civil and Environmental Engineering, Stanford University, Stanford, CA 94305-4020, USA, E-mail: borja@stanford.edu

Author

Magued Iskander, PhD, PE, F.ASCE
Civil Engineering Department
Polytechnic Institute of New York University
Six Metrotech Center
Brooklyn, NY 11201
USA
E-mail: Iskander@poly.edu

ISBN 978-3-642-13107-3 e-ISBN 978-3-642-13108-0

DOI 10.1007/978-3-642-13108-0

Springer Series in Geomechanics and Geoengineering ISSN 1866-8755

Type Design and Cover Design: Scientific Publishing Services Pvt. Ltd., Chennai, India.

Printed on acid-free paper

5 4 3 2 1 0

springer.com

To

Sherry,

Lolli, Kito,

& My Parents

Contents

Acknowledgements

Most of the ideas presented in this manuscript originated at a time when I worked with Professor Roy Olson at The University of Texas at Austin. I am grateful to Professor Olson for his advice, friendship, and his careful review of an earlier version of this manuscript. It has been both a privilege and a pleasure to have worked with him.

A large number of professional colleagues contributed formally and informally to the success of this effort. Professors James Bay, Utah State University; J. Don Murff, Texas A&M University; Ken Stokoe, The University of Texas at Austin; Priscilla Nelson, New Jersey Institute of Technology; Jean-Louis Briaud, Texas A&M University; Jack Templeton, Sage Engineering; and the Late Professors Lymon Reese, The University of Texas and Michael O'Neill, The University of Houston provided valuable advice and crucial insights. Aaron Goldberg, PE; Sanjeev Malhotra, PE; Casey Jones, PE; James Doran; Robert Pavlicheck, PE; and Dr. Omar Alansari assisted in equipment design, manufacturing, and testing. My graduate students Saumil Parikh, and Bhujang Patel proofread the entire manuscript.

Portions of this research were funded by NSF's Offshore Technology Research Center (Grant No. CDR 8721512). Continued NSF funding under grants No. CMMI 9733064, DGE 0337668, and DGE 0741714 is gratefully acknowledged.

I am grateful to Dr. Thomas Ditzinger for adopting this manuscript into the Springer Series on Geomechanichs and Geoengineering. I am also thankful to the series editors: Prof. Wei Wu from Universität für Bodenkultur in Austria and Prof. Ronaldo Borja from Stanford University for recommending publication of this manuscript. Credit is also due to Mrs. Heather King from Springer for her assistance during the production of this work.

Last but not least, I would like to express my gratitude to my family, particularly my parents and my wife. This work could not have been finished without their support, encouragement, and love.

Selected Unit Conversions

Conversions from American (English) to SI Units

Length	1 in. = 2.54 cm
	1 ft = 0.3048 m
Area	1 in.^{2} = 6.452 cm^2
	1ft^2 = 929.03 cm^2
Volume	1 in.^{3} = 16.387 cm^3
	1 ft^3 = 28.317 x10^{-3} m^3
Force	1 lb = 4.448 N
	1 kip = 4.448 kN
	1 US Ton = 8.896 kN
Stress	1 psi = 6.895 kN/m^2
	1 psf = 47.88 N/m^2
	1 tsf = 95.76 kN/m^2
	1 kip/ft^2 = 47.88 kN/m^2
Unit Weight	1 pcf = 0.1572 kN/m^3
Moment	1 lb ft = 1.3558 N.m
	1 lb-in. = 0.11298 N.m
Energy	1 ft-lb = 1.3558 J

Conversions from SI Units to American (English) Units

Length	1 m = 3.281 ft
	1 cm = 0.3937 in
Area	1 m^2 = 10.764 ft^2
	1 cm^2 = 0.155 in.^{2}
Volume	1 m^3 = 35.32 ft^3
	1 cm^3 = 0.061023 in.^{3}

Force	1 N = 0.2248 lb
	1 kN = 0.2248 kip
	1 metric ton = 2204.6 lb
Stress	1 kN/m^2 = 20.885 psf
	1 kN/m^2 = 0.01044 tsf
	1 kN/m^2 = 0.145 psi
Unit Weight	1 kN/m^3 = 6.361 pcf
Moment	1 N.m = 0.7375 lb-ft
	1 N.m = 8.851 lb-in.
Energy	1 J = 0.7375 ft-lb

1. Introduction

Abstract. Design guidelines for predicting the axial capacity of piles in sand have long been the source of considerable debate due to the geotechnical profession's inadequate understanding of pile behavior. Analytical studies, alone, cannot resolve the uncertainties in axial capacity predictions due to the complex nature of pile behavior in sand. This chapter presents some of the factors which occur during installation and loading of pipe piles in sand, along with their potential impact on axial capacity.

1.1 Background

There are many challenges facing the designers of pile foundations in sand, particularly in offshore environments. Most offshore structures are supported on steel pipe piles which are designed in accord with *API Recommended Practice for Planning, Designing, and Constructing Fixed Offshore Platforms, RP-2A* (API, 2000). The recommended practice has long been the accepted standard for pile design as evidenced by its wide spread use (Lacasse 1988, Pelletier *et al.* 1993). The document, RP-2A, is currently in its twenty first edition, however, little has changed in the recommended practice for piles in sand since the fifteenth edition was published in 1984. The publication of the fifteenth edition, with significant changes in the criteria for predicting the shaft capacity of piles in sand was controversial and, instigated a round of criticism and debate (*e.g.* Lings, 1985; Randolph, 1985; and Toolan *et al.* 1990).

The accuracy of the API approach has been evaluated by comparing the predicted pile capacities to the measured pile capacities of known case histories (Olson and Iskander 1994, Iskander and Olson 1992). The scatter between measured and predicted pile capacities is significantly larger for piles in sand than piles in clay (Olson and Dennis, 1982). The primary inadequacy of the API method is the database used to calibrate the design method. At the present time, the database has fewer load tests than the number of degrees of freedom of the pile-soil system, which precludes determining a unique set of calibration factors (Olson and Iskander 2009). Furthermore, the database consists of piles that are generally shorter in length, smaller in diameter, and have less capacity than the piles being designed for offshore structures.

Variations between measured and predicted pile capacities arise due to a number of factors such as severe changes in soil properties and state of stress caused by pile driving, difficulties in measuring field soil properties particularly in the marine environment, variations in loading details and pile installation procedures,

M. Iskander: Behavior of Pipe Piles in Sand, SSGG, pp. 1–6.

uncertainties in the behavior of the soil core (plug) during driving and subsequent loading, and soil structure interaction. Most design methods, including the API method, account for these factors by calibrating a simple theoretical model using empirical factors.

Although these factors are most pronounced in offshore piling, they are equally relevant to traditional terrestrial geotechnical engineering. In the terrestrial environment, many uncertainties in pile capacity predictions can be avoided by driving piles to "refusal" in deep layers of hard clay, dense sand, or rock. When strong layers are not available, and driving piles to refusal is not possible, engineers often apply conservative assumptions and rely on proof tests to verify the design capacity. So a better understanding of the effects of installation on the capacity of piles has the potential to provide cost savings for both terrestrial and offshore piles.

1.2 Installation Effects That Influence the Capacity of Pipe Piles in Sand

1.2.1 Plugging of Piles

During installation of open-ended pipe piles, the soil enters the pile at a rate that is equal to, or larger than, the rate of pile penetration. This mode of penetration is referred to as *coring* or *cookie cutter*. As penetration progresses, the soil core inside the pile may develop sufficient frictional resistance along the inner pile wall to prevent further soil intrusion, causing the pile to become *plugged*. Plugging is important, not only because it directly contributes to the tip bearing capacity, but also because it indirectly contributes to the developed shaft capacity (Gavin and Lehane 2002, Paik and Salgado 2003), since a plugged pile displaces more soil than a coring one, which increases the effective stresses surrounding the pile. Plugging also influences the dynamic behavior of piles, which complicates the dynamic analyses of piles (Paikowsky and Whitman, 1990; Raines *et al.* 1992).

The common view is that most piles that plug during static loading do not plug during driving (Paik and Lee 1993, Hight *et al.* 1996, Hajduk *et al.* 2009). Smith and Chow (1982) attributed this phenomenon to a combination of inertial effects and an increase in the dynamic bearing capacity factor, Nq, over its static value. Additionally, Smith *et al.* (1986) argued that the driven pile mobilizes all of its internal and external friction intermittently during penetration and, as a result, the soil core advances up the pile. Conversely, Paikowsky *et al.* (1989) argued that "the pile plugging phenomenon is of frequent occurrence and is of greater significance than that presently accorded it by the profession." The degree of plugging depends on soil properties (Lehane and Gavin 2001), the hammer characteristics, and the pile's frictional resistance (Brucy *et al.* 1991).

On some occasions, piles may plug and impede driving. If the available pile hammer cannot drive the pile to the design depth a problem may arise, particularly for piles with thickened walls near the surface or mud-line, such as piles used to resist lateral loading (Murff *et al.*, 1990). If the pile "refuses" prematurely, the required thick section may end up above the mud-line. In these circumstances, the plug is typically removed by drilling or jetting. The effects of the removal of the soil plug on the final pile capacity are controversial. Brucy *et al.* (1991) claimed that static loading results are not changed by partial removal of the soil plug. Other studies indicate that jetting results in a substantial reduction in the over-all pile capacity, even if the pile is re-driven (McClelland and Focht, 1955).

Arching of soil within a few pile diameters from the tip is sufficient to form a plug (Leong and Randolph 1991). Furthermore, the available soil friction along the inside of pipe piles is significantly higher than the common design values for the external skin friction, for sands (Paik *et al.* 2003, Murff *et al.* 1990), clays (Randolph, 1987), and even for calcareous soils (Randolph *et al.* 1990). Small deformations on the order of 2 mm are required to mobilize arching that is necessary for plugging, particularly in dense sands (Kishida *et al.* 1985). Nevertheless, piles that plug during static loading may exhibit smaller tip bearing capacities than their closed-end counterparts at the same pile head settlement (Szechy 1958, Briaud and Audibert 1990). O'Neill and Raines (1991) concluded that the reduction in tip bearing capacity is primarily due to the plug's compressibility, while Lehane and Randolph (2002) ascribe the effect to compressibility of the soil below the plug.

De Nicola and Randolph (1997) found that the plug length increased with increasing relative density of insitu sands during driving and decreased with increasing relative density during jacking. Also, during installation, the jacked piles exhibited a greater tendency to plug than the driven piles. This points to the importance of testing driven model piles.

1.2.2 Buildup of Pore Water Pressure

Arching of soils inside the pile to form a plug depends on the ability of the soil to drain pore pressures and develop high frictional stresses along the soil/pile interface. Under cyclic or earthquake loading, the soil core may become partially drained, which may prevent pile plugging (Randolph *et al.* 1991), or reduce pile capacity (Choi and O'Neill 1997).

There has been increasing recognition of gain in pile capacities in sand with time. York *et al.* (1994) attributed the gain over a period of a few weeks to dissipation of pore water pressure. Chow *et al.* (1998) attributed the gain over a period of 5 years to creep that leads to break down of circumferential arching stresses allowing increase in radial stress as well as increased dilatation due to aging.

1.3 Scope and Objectives

Analytical studies, alone, cannot resolve the uncertainties in axial capacity predictions due to the complex nature of pile behavior. Experimental research on the physical processes that control the behavior of piles during installation and loading is required in order to resolve the issues involved. Considering the large number of variables involved in axial pile capacity development, as well as the high cost of load tests on full scale instrumented piles, it is unlikely that a sufficient number of full scale load tests can be performed to resolve the problem. Furthermore, the test conditions of full-scale load tests, generally, cannot be controlled satisfactorily to isolate independent variables. Experimental research on model piles can, however, be used to resolve some of the important design uncertainties.

The objectives of this monograph presents a detailed blue print for developing experimental facilities necessary to investigate the effects of plugging and pore-water pressure generation on the capacity of piles. These facilities include a unique instrumented double-walled pipe pile that is used to delineate the frictional stresses acting against the external and internal surfaces of the pile. The pile is fitted with miniature pore-pressure transducers to monitor the generation of pore water pressure during installation and loading. A fast automatic laboratory pile hammer capable of representing the phenomena that occur during pile driving was also developed and used. Finally, a pressure chamber; feedback control system; data acquisition system; loading frame; sand handling, pluviating, saturating, and drying apparatus have been integrated to allow convenient load testing of piles under simulated field conditions. The experimental apparatus is presented with sufficient details to allow readers to duplicate or modify the design to suit their own needs.

A number of load tests were carried out to identify the effects of inertia and build-up of pore water pressure on pile plugging. Continuous measurement of dynamic and static excess pore pressures, frictional and end bearing stresses, and the elevation of the soil inside the pile during installation and loading are presented. The results of the testing program validate the performance of the developed apparatus, and provide unique insights into soil-structure interaction during pile driving and subsequent loading. The work contributes to a better understanding of pile behavior.

1.4 Organization of This Book

This monograph consists of four main parts. The subject is introduced along with previous research in chapters one, two, and three. The design and performance of the experimental facilities is presented in chapters four, five, and six. The properties of the test soil and scaling considerations are presented in chapters seven and eight, respectively. A testing program to investigate the behavior of piles during installation and subsequent loading is presented in chapter nine. A summary of the work along with conclusions are presented at the end.

References

American Petroleum Institute (API): Recommended Practice for Planning, Designing, and Constructing Fixed Offshore Platforms, Working Stress Design, 21st edn. (RP 2A-WSD), American Petroleum Institute (2000)

Briaud, J., Audibert, J.: Some Thoughts on API RP2A for Vertically Loaded Piles. In: Proc. 22nd Offshore Technology Conf., pp. 9–16 (1990), doi:10.4043/6418-MS

Brucy, F., Meunier, J., Nauroy, J.F.: Behavior of Pile Plug in Sandy Soils During and After Driving. In: Proc. Offshore Technology Conf., Paper No: 6514-MS (1991), doi:10.4043/6514-MS

Choi, Y., O'Neil, M.: Soil Plugging and Relaxation in Pipe Pile During Earthquake Motion. ASCE J. Geotechnical and Geoenvironmental Engr. 123(10), 975–982 (1997)

Chow, F., Jardine, R., Brucy, F., Nauroy, J.: Effect of Time on Capacity of Pipe Piles in Dense Marine Sand. ASCE J. Geotechnical and Geoenvironmental Engr. 124(3), 254–264 (1998)

De Nicola, A., Randolph, M.: The Plugging Behaviour of Driven and Jacked Piles in Sand. Geotechnique 47(4), 841–856 (1997)

Gavin, K., Lehane, B.: The shaft capacity of pipe piles in sand. Canadian Geotechnical J. 40, 36–45 (2002)

Hajduk, E., Lin, G., Adams, J., Ledford, D.: Experiences with Open Ended Pipe Pile Plugging in The Atlantic Coastal Plain. In: Iskander, M., et al. (eds.) Contemporary Topics in Deep Foundations (GSP 158), pp. 1–10 (2009), doi:10.1061/40902(221)38

Hight, D., Lawrence, D., Farquhar, G., Mulligan, G., Gue, S., Potts, D.: Evidence for Scale Effects in the End Bearing Capacity of Open-Ended Piles in Sand, Paper No. 7975-MS (1996), doi:10.4043/7975-MS

Iskander, M.G., Olson, R.E.: Review of API Guidelines for Pipe Piles in Sand. In: Proc., Civil Engineering in the Oceans V, ASCE, pp. 798–812 (1992)

Kishida, H., Isemoto, N.: Behavior of Sand Plugs in Open Ended Steel Pipe Piles. In: Proc. 9th Int. Conf. on Soil Mechanics and Foundation Engineering, Tokyo, pp. 601–604 (1977)

Kishida, H., Uesugi, M., Susumu, M.: Behavior of Dry Sands in Steel Pipe Piles. In: Proc. 8th Southeast Asian Geotechnical Conf., Kuala Lumpur (1985)

Lacasse, S.: Uncertainties in Offshore Geotechnical Engineering, Int. Survey of API-RP 2A Design Parameters for Axial Capacity of Driven Piles in Sand, Norwegian Geotechnical Institute, Report No: 85307-14 (1998)

Lehane, B., Gavin, K.: Base Resistance of Jacked Pipe Piles in Sand. ASCE J. Geotechnical and Geoenvironmental Engineering 127(6), 473–480 (2001)

Lehane, B., Randolph, M.: Evaluation of a Minimum Base Resistance for Driven Pipe Piles in Siliceous Sand. ASCE J. Geotechnical and Geoenvironmental Engineering 128(3), 198–205 (2002)

Leong, E., Randolp, M.: h Finite Element Analysis of Soil Plug Response. Int. Journal of Numerical and Analytical Methods in Geomechanics 15, 121–141 (1991)

Lings, M.L.: The Skin Friction of Driven Piles in Sand, M.Sc. Thesis, Imperial College of Science and Technology, University of London (1985)

McClelland, B., Focht, J.: Padre Island Load Tests, Project Report, McClelland Engineers (Fugro-McClelland Marine Geosciences), 6100 Hilcroft, Houston, Texas (1955)

Murff, J., Raines, R., Randolph, M.: Soil Plug Behavior of Piles in Sand. In: Proc. 22nd Offshore Technology Conf., Houston, TX, pp. 25–32 (1990)

Murff, J.D.: Axial Pile Capacity in Sand, Draft Proposal for API 1990 Research Project, Exxon Production Research Company, P.O. Box 2189, Houston, Texas 77252 (1998)
O'Neill, M.W., Raines, R.D.: Load Transfer for Pipe Piles in Highly Pressured Dense Sand. Journal of Geotechnical Engineering, 1208–1226 (1991)
Olson, R., Iskander, M.: Axial Load Capacity of Un-Tapered Piles in Cohesionless Soils. In: Contemporary Topics in Deep Foundations. GSP No. 185, pp. 231–238. ASCE (2009)
Olson, R.E., Dennis, N.D.: Review and Compilation of Pile Load Test Results, Axial Pile Capacity, Final Report to The American Petroleum Institute on Project PRAC 81-29, Geotechnical Engineering Report CR83-4. Department of Civil Engineering, The University of Texas at Austin (1982)
Olson, R.E., Iskander, M.G.: Axial Load Capacity of Pipe Piles in Sands. In: Proc. 1st Regional Conf. and Exhibition on Advanced Technology in Civil Engineering, Manama, Bahrain, pp. 383–394. ASCE–SAS (1994)
Paik, K., Salgado, R.: Determination of Bearing Capacity of Open-Ended Piles in Sand. ASCE J. Geotechnical & Geoenvironmental Engr. 129(1), 46–57 (2003)
Paik, K., Salgado, R., Lee, J., Kim, B.: Behavior of Open- and Closed-Ended Piles Driven Into Sands. ASCE J. Geotechnical & Geoenvironmental Engr. 129(4), 296–306 (2003)
Paik, K.H., Lee, D.R.: Behavior of Soil Plugs in Open Ended Model Piles Driven Into Sands. Marine Geotechnology 11, 353–373 (1993)
Paikowsky, S., Whitman, R.: The Effects of Plugging on Pile Performance and Design. Canadian Geotechnical J. 27(4), 429–440 (1990)
Paikowsky, S., Whitman, R., Baligh, M.: A New Look at The Phenomena of Offshore Pile Plugging. Marine Geotechnology 8, 213–230 (1989)
Pelletier, J.H., Murff, J.D., Young, A.C.: Historical Development and Assessment of the Current api Design Methods for Axially Loaded Pipes. In: Proc. Offshore Technology Conf., Houston, TX, OTC paper 7157, pp. 253–282 (1993)
Raines, R.D., Ugaz, O., O'Neill, M.: Driving Characteristics of Open Towe Piles in Sand. Journal of Geotechnical Engineering, 72–88 (1992)
Randolph, M.: Modeling of The Soil Plug Response During Pile Driving. In: Proc. 9th Asian Geotechnical Conf., Bangkok, Thailand, pp. 6-1–6-14 (1987)
Randolph, M., Leong, E., Hyden, A., Murff, J.: Soil Plug Response in Open Ended Pipe Piles. ASCE J. of Geotechnical & Geoenvironmental Engr., 743–759 (1990)
Randolph, M.F., Leong, E.C., Houlsby, G.T.: One Dimensional Analysis of Soil Plugs in Pipe Piles. Geotechnique 41, 587–598 (1991)
Randolph, M.F.: Capacities of Piles Driven into Dense Sands. Presentation to the 11th Conf. of Soil Mechanics and Foundation Engineering, San Francisco, Session 4C (1985)
Smith, I., Chow, Y.: Three Dimensional Analysis of Pile Drivability. In: Proc. 2nd Int. Conf. on Numerical Methods in Offshore Piling, Austin, Texas, pp. 1–19 (1982)
Smith, I., To, P., Wilson, S.: Plugging of Pipe Piles. In: Proc. 3rd Int. Conf. on Numerical Methods in Offshore Piling, Nantes, France, pp. 54–73 (1986)
Szechy, C.: Tests with Tubular Piles. Acta Technica, Hungarian Academy of Science 24, 181–219 (1958)
Toolan, F.E., Lings, M.L., Mirza, U.A.: An Appraisal of api rp-2a Recommendations for Determining Skin Friction of Piles in Sand. In: Proc. 22nd Offshore Technology Conf., Houston, TX, OTC paper 6422 (1990)
York, D., Brusey, W., Clemente, F., Law, S.: Setup and Relaxation in Glacial Sand. ASCE J. Geotechnical Engr. 120(9), 1496–1513 (1994)

2. Review of Design Guidelines for Piles in Sand

Abstract. This chapter examines the current API RP-2A design guidelines for piles in sand in the light of available load test data. The main weaknesses of the design approach are also identified.

2.1 Introduction

Pile foundations have been used for many years, and as a result a large number of design approaches have been proposed to predict their capacity. These approaches range from simple empirical approaches to sophisticated dynamic finite element analyses, with new methods introduced every few years. Darrag (1987) summarized most of the widely recognized design methods.

Empirical pile design methods, where pile capacity is correlated to in situ tests, have been popular, particularly in Europe. The most widely used methods are based on the cone penetration test (e.g. Notingham 1975), the standard penetration test (e.g. Meyerhof 1976), and the pressuremeter test (e.g. Sellgren 1982). In situ tests reflect, to some extent, field conditions but fail to model and/or measure some of the important parameters that influence pile behavior. Thus, empirical methods often give good predictions when correlated with load test results on a regional basis, but tend not to excel as a generalized prediction method.

Analytical methods based on simplified limit equilibrium or cavity expansion models have also been used to predict pile capacity (*e.g.* Vesic 1967 and 1972). Analytical methods typically use design parameters which describe soil conditions and stresses prior to pile installation, and do not account explicitly for the substantial change in soil conditions and state of stress due to pile driving. Additionally, analytical methods do not account for soil compressibility and soil structure interaction. At this stage, analytical methods are most useful in preliminary computations, as well as in extrapolating load test results to other pile dimensions at the same site.

Experience-based design methods such as the API method (American Petroleum Institute Recommended Practice RP-2A, 2000) use an analytical framework, but are calibrated using load test results. Thus, experience-based methods should be considered empirical and are used mainly as a scheme for generalizing the results of pile load tests to new design situations. Studies of pile load test databases have shown that the margin of error in the static axial load capacities of piles predicted using the API method is larger for piles in cohesive soils than for comparable piles in cohesive soils (Dennis and Olson 1983a and 1983b). The margin of error

M. Iskander: Behavior of Pipe Piles in Sand, SSGG, pp. 7–23.
springerlink.com

in the prediction is particularly high for open-ended steel pipe piles, such as the piles used to support offshore oil production platforms (Olson 1990; Toolan *et al.* 1990, Hossain and Briaud 1992, Olson and Iskander 1994).

In recent years gains in computing power have made realistic Finite Element Method simulations possible (El-Mabsout 1991). Although FEM analysis are popular among practicing engineers, they suffer from either use of simplified soil models or when using advanced constitutive models the need for a large number of soil parameters that are not typically measured by practicing engineers. The cost of measuring the required soil properties may be prohibitive. Additionally, few analyses have been supported by full scale field load tests, and thus the accuracy of the final result is usually unsubstantiated. The existence of these, and other problems, provides a challenge for researchers but, currently, does not provide practicing engineers with a reasonable design tool, for pipe piles in sand.

Experience-based methods, such as the API method, are, by far, the most widely used methods to predict the capacity of piles. Two surveys of offshore engineers (Lacasse and Goulois 1989; Focht and O'Neill 1985) revealed that most designers follow the API recommended practice (API RP-2A 2000 section 6.4.3). The wide use of the API method has initiated considerable debate among researchers and practicing engineers regarding the accuracy of the pile capacities predicted using RP-2A. This debate has motivated the American Petroleum Institute (API), The National Science Foundation (NSF) Offshore Technology Center (OTRC), The British Department of Energy (DOE), several oil companies, among others, to sponsor a number of research studies in an attempt to resolve some of the design uncertainties. Results of these studies are inconclusive mainly due to the paucity of load test data, particularly for large piles, as well as the poor quality of most of the available load test data.

This chapter examines the API RP-2A (2000) design guidelines for piles in sand in the light of available load test data. The main weaknesses of the API method are also identified. The effects of the installation process on the axial capacity of pipe piles in sand are discussed in detail in chapter three.

2.2 API Design Guidelines for Piles in Sand

The axial load capacity (Q_C) of a driven pipe pile in cohesionless soils is estimated using API (2000) as follows:

$$Q_C = Q_S + Q_P \pm W_P \tag{2.1}$$

where:

Q_S capacity in side shear

Q_P tip capacity, taken as zero for piles in tension

Wp weight of the pile submerged in soil, positive for tensile loading and negative for compressive loading. Wp is often neglected due to it's small contribution.

The side shear, Q_S, is given by:

$$Q_S = \sum_{i=1}^{N} f_s C \Delta L \tag{2.2}$$

$$f_S = K\,\overline{\sigma}_{vo}\,tan(\delta) \leq f_{lim} \tag{2.3}$$

where:

- f_S local side shear between the pile and the surrounding soil, limited to f_{lim}
- C pile circumference
- ΔL increment of pile length in the $i'th$ layer
- $\overline{\sigma}_{vo}$ free field vertical effective stress in the middle of the $i'th$ layer
- K ratio of horizontal to vertical effective stresses
- δ pile/soil friction angle
- f_{lim} limiting skin friction stress

The tip capacity, Q_P, is given by:

$$Q_P = q_P\ A_P \tag{2.4}$$

$$q_P = \overline{\sigma}_{vo}\ N_q \leq q_{lim} \tag{2.5}$$

where:

- q_P net pressure between the pile tip and soil limited to q_{lim}
- A_P tip area of the pile
- N_q dimensionless bearing capacity factor
- $\overline{\sigma}_{vo}$ free field vertical effective stress at the elevation of the pile tip
- q_{lim} limiting end bearing stress

For open-ended steel pipe piles, the tip capacity is the smaller of: (1) the tip bearing capacity of an equivalent closed-ended pile, or (2) the end bearing on the steel rim plus the side shear capacity of the soil core inside the pile, computed using Eq. 2.2 and 2.3.

The properties currently recommended by API are shown in Table 2.1. The earth pressure coefficient, K, is assumed to be 1.0 for closed-ended piles and 0.8 for open-ended piles, with tension and compression treated the same.

This form of the analysis has been used by API for a number of years, but the current values are based on a database of load tests prepared by Olson and Dennis (1982).

Table 2.1 API RP-2A Design Values for Cohesionless Soils

Relative Density	Soil Description	δ (deg.)	f_{lim} (ksf)	N_q	q_{lim} (ksf)
very loose	sand	15	1.0	8	40
loose	sand-silt				
medium	silt				
loose	sand	20	1.4	12	60
medium	sand-silt				
dense	silt				
medium	sand	25	1.7	20	100
dense	sand-silt				
dense	sand	30	2.0	40	200
very dense	sand-silt				
dense	gravel	35	2.4	50	250
very dense	sand				

2.3 The Database

A database of about 1000 load tests on piles of various types, and in a variety of soils, was developed during the early 1980's, and a method for analysis of pipe piles in sand was recommended (Olson and Dennis 1982, Dennis and Olson 1983b). Later, efforts were made to expand the size of the database and to improve its interpretation (Olson and Al-Shafei 1988, Olson 1990). This effort has continued in Iskander and Olson (1992) and Olson and Iskander (1994, 2009) and is the basis for the analyses presented herein.

Despite the large size of the database, there are relatively few load tests on steel pipe piles in sand. Olson (1990) used data for only four open-ended pipe piles in sand along with data for 31 closed-ended pipe piles in sand. The current study uses a database of 47 load tests, of which thirteen are for open-ended pipe piles (Table 2.2). The piles used in this study ranged in length from 10 to 140 ft, and in diameter from 8 to 32 inches.

In applying the API recommended practice to predict the capacity of the piles in the database, it is necessary to adopt a standard method to define relative density. Since API does not provide guidelines for determining relative density, the classification recommended by Peck *et al.* (1974) was followed. Relative density was defined using standard penetration resistance, N, in blows per foot (Table 2.3). The N values were corrected for the effects of overburden pressure using the recommendation of Peck *et al.* (1974) as follows:

$$N_{Corrected} = C_N \cdot N_{Measured} \tag{2.6}$$

$$C_N = 0.77 \ Log \frac{20}{\overline{\sigma}_{vo}} \tag{2.7}$$

where:

C_N correction factor valid for $\sigma_{vo} \geq 0.25$ tons/ ft^2

$\overline{\sigma}_{vo}$ vertical effective overburden pressure in tons/ ft^2 at the elevation of the Standard Penetration Test

Generally, piles in sand do not experience a sudden "plunging" failure. Several definitions of "failure" can be adopted under these circumstances. For the analyses reported herein, failure was defined as the peak applied load during the test, but tests were not accepted unless loads reached at least the value defined by Davisson (1973) which corresponds to the following pile head settlement:

$$S = \frac{QL}{AE} + 0.15'' + 0.1D_b \tag{2.8}$$

where:

S pile head settlement corresponding to the ultimate load
QL/AE elastic compression of pile head in inches
Q load in pounds
L pile length in inches
A cross sectional area of the pile in square inches
E modulus of elasticity of pile material in psi
0.15 displacement in inches required to mobilize side shear
$0.1D_b$ displacement in inches required to mobilize end bearing
D_b diameter of the pile in feet

Table 2.2 Database of Load Tests on Steel Pipe Piles

Load Test No.	Location	Length (ft)	Comp.\ Tens.	DQF	Q_m (kips)	Q_c (kips)
89	Arkansas River	53	C	3	360.0	351.1
90	Arkansas River	53	T	3	180.0	197.3
95	Arkansas River	53	C	3	560.0	572.2
96	Arkansas River	53	T	3	230.0	222.5
102	Arkansas River	53	C	3	500.0	415.8
103	Arkansas River	53	T	2	240.0	228.8
112	Old River	66	C	3	750.0	827.2
114	Old River	66	C	3	750.0	602.9
115	Old River	45	C	3	280.0	178.0
116	Old River	65	C	3	700.0	702.0
117	Old River	65	C	3	680.0	656.3
118	Old River	65	T	3	390.0	316.8
120	Old River	66	T	3	400.0	268.0
122	Old River	65	T	3	370.0	288.5
143	Ogeechee River	10	C	5	200.o	26.3
144	Ogeechee River	20	C	5	480.0	92.5
145	Ogeechee River	29	C	5	630.0	152.4
146	Ogeechee River	39	C	5	800.0	214.8
147	Ogeechee River	49	C	5	860.0	411.9
148	Ogeechee River	49	T	5	346.0	195.4
205	Helena, Arkansas	80	C	2	285.0	424.5
206	Helena, Arkansas	55	C	2	136.0	246.9
210	Helena, Arkansas	60	C	2	210.0	272.6
238	San Francisco	30	C	5	113.0	79.2
243	Tokyo, Japan	27	T	2	370.0	133.3
244	Tokyo, Japan	13	C	2	68.0	28.0
245	Tokyo, Japan	34	C	2	310.0	120.0
341	Kansas City	55	C	3	188.0	252.6
342	Kansas City	55	C	3	218.0	285.9
353	Mustang Island, TX	69	T	4	455.0	375.2
518	St. Luis	37	T	2	80.0	65.5
519	St. Luis	44	T	2	90.0	100.6
722	Jackson, MS	28	C	2	280.0	60.7
725	Mobile	56	C	2	270.0	94.6
796	Muskegon, MI	58	C	3	240.0	169.5
804	Louisville, KY	70	T	2	110.o	375.9
805	Louisville, KY	70	C	2	280.0	564.2

Table 2.2 (*continued*)

Load Test No.	Location	Length (ft)	Comp.\ Tens.	DQF	*Qm* (kips)	*Qc* (kips)
812	Confidential (TX)	48	T	4	122.0	177.4
813	Confidential (TX)	56	T	4	166.0	257.9
934	New Haven	130	C	3	240.0	726.1
952	Winnemucca, NV	58	C	3	242.0	473.1
1021	Ras Tanajib	59	T	4	2850.0	519.0
2000	Confidential (SA)	69	T	1	1108.8	382.7
2001	Confidential (SA)	41	T	1	1248.7	266.2
9064	Japan	97	C	2	719.2	2299.1
9141	Japan	130	C	2	535.42	4215.7
9142	Japan	138	C	2	659.29	4470.9

DQF: Data quality factor ranging from 1–5, with larger numbers indicating better data.

Table 2.3 Standard Definition of Relative Density

Relative Density	Blows per Foot
Very Loose	0 – 4
Loose	4 – 10
Medium Dense	10 – 30
Dense	30 – 50
Very Dense	over 50

Table 2.4 Standard Definition of Unit Weights

Soil Description	Unit Weight, pcf
Gravel	130
Sand Gravel Mixtures	130
Clean Sand	125
Silty Sand	120
Sandy Silt	120
Silt	115

For cohesionless soils, essentially no measured unit weights were reported and there are no known reliable correlations between unit weights and blow counts (N values). Accordingly, unit weights were standardized on the values given in Table 2.4.

An examination of the existing database discloses a number of limitations (Iskander and Olson, 1992). The data are largely for piles in medium to dense clean sands. For side shear, 63% of the capacity is estimated to come from clean sand, 28% from silty sands and sandy silts, 4% from silt, 3% from sand/gravel mixtures, and 1% from clay. For the tip capacity, 89% is estimated to come from clean sand, 8% from sand/gravel mixtures, 1% from gravel, 2% from sand/silt mixtures, and none from silt or clay. In the offshore, silty soils are not well represented because, geologically, the finer soils tend to occur at shallow depths. The low effective stresses at shallow depths lead to reduced frictional stresses. Piles are typically driven to firm material and thus logically tend to derive most of their side shear and tip capacities from deeper, thus coarser, soils. Significant vertical and lateral variations in the properties of cohesionless soils may also result in soil borings that do not accurately reflect the soil properties at the location of the load test. When blow counts are used for soil classification, variations in the efficiency of the hammers could also lead to scatter, especially when testing methods differ with respect to time or regionally.

2.4 Analysis Using the 2000 API RP 2A

Analyses were performed for the piles in the database using the 1993 API RP-2A recommendations. Measured and predicted (calculated) capacities are compared in Fig. 2.1, with hollow symbols used for tests in which the quality of the source data was judged to be deficient, and the solid symbols for cases involving better data. The ratio of calculated-to-measured axial load capacity (Q_c/Q_m) ranged from 0.13 to 3.57. The ratios of Q_c/Q_m tend to be normally distributed when $\log(Q_c/Q_m)$ is used. The mean $\log(Qc/Qm)$ was -0.129 (antilog is 0.74), which indicates that the API method tends to under-predict capacities. The large scatter indicates that the method can be unsafe in some cases and uneconomical in others.

Use of the API method led to under-prediction of the capacities of short piles and over-prediction of the capacity of long piles (Fig. 2.2). The over-predictions may be large for piles of lengths often used offshore.

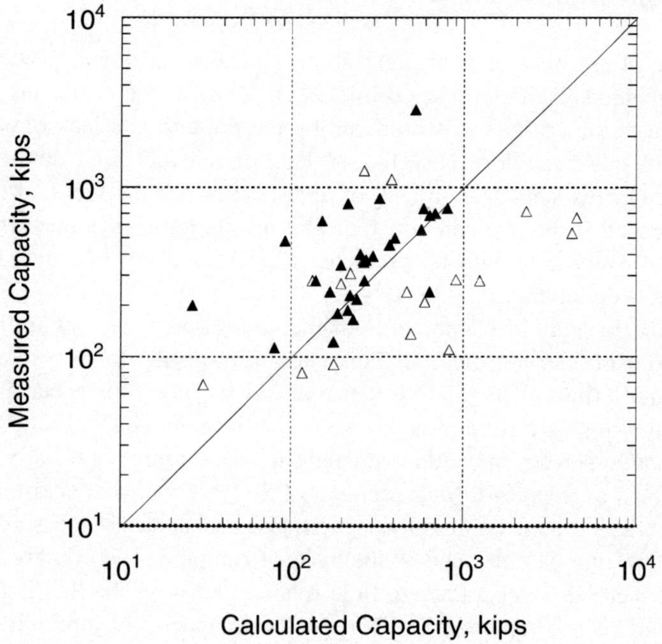

Fig. 2.1 Comparison of Measured Capacities to Capacities Predicted Using API RP-2A

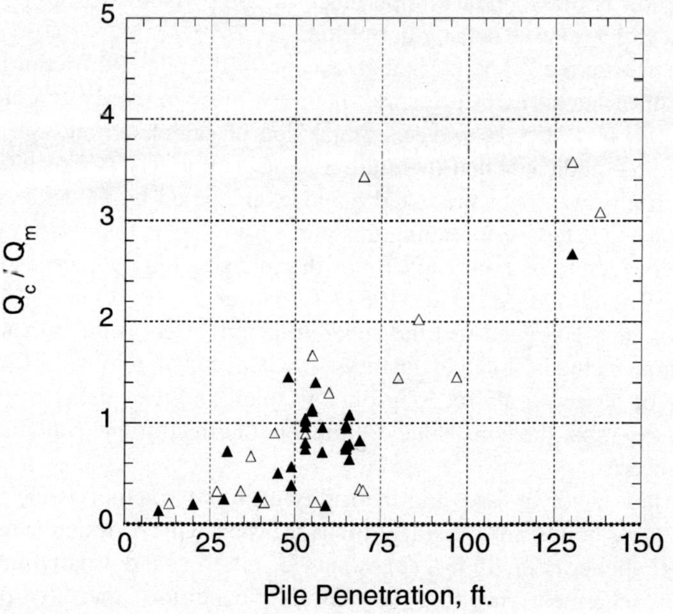

Fig. 2.2 Effect of Penetration on Capacities Computed with API RP-2A

2.5 Assessment of the API Method for Piles in Sand

The wide use of API RP-2A has brought about numerous criticism as well as support for the method (Iskander and Olson 1992). Many of the criticisms are valid, but resolution of the criticisms is hindered by the profession's lack of a good understanding of pile behavior, as well as a lack of relevant high quality data. A number of criticisms stem from a lack of appreciation for the fact that the guidelines are intended to be a prediction tool and are therefore not intended to be a lower bound "safety net" with all predicted capacities being less than the measured capacities (Pelletier *et al.* 1993).

Hossain and Briaud (1992) pointed out the large user variability in the capacities predicted using the API method. This variability results for two reasons; lack of an accurate definition of relative density, and the need for a clarification on whether the limiting skin friction is intended as a limit on the local or average skin friction. Relative density of sands is difficult to measure and is usually estimated from blow count correlations. Measurement of SPT blow counts is rare in the offshore environment; more commonly, penetration resistance is measured using a down-hole wire-line hammer and a Shelby tube sampler. Accordingly, engineers use different methods to characterize field relative density. The limiting skin frictions in API RP-2A recommendations for the axial capacity of pipe piles are typically used as local values. Lings (1985), Vesic (1967), and others, have applied limiting stresses in side shear to an average side shear over the entire depth of the pile. Hossain and Briaud (1992) argued, however, that the definition of the limiting skin friction is of secondary importance in comparison with user variability due to the lack of a relative density definition.

The consensus of the "experts" is that, except for the pile-soil friction angle and the coefficient of lateral earth pressure, the guidelines are overly conservative by 20–40% (Lacasse 1988). However, comparison of calculated and measured capacities (Fig. 2.1) indicates that there are a number of case histories in which the axial pile capacity was over predicted or under predicted by values in excess of 40%. A number of studies, including this one, also indicate that the API method is likely to over-predict the axial capacity as the pile penetration exceeds 60 ft (*e.g.* Lings 1985, Briaud and Anderson 1987, Toolan *et al.* 1990, and Olson 1990). Pelletier *et al.* (1993) argued that the apparent length effect is due to variations in the installation methods, loading history, soil density, or removal of soil plugs. The number of observations for long piles is small and the quality of the data is generally suspect, but the consistency of the length effect in the available data is a cause for concern.

An important factor in assessing the safety of the API method is the realization that (1) few data points are considered unsafe when typical design safety factors are applied (Pelletier *et al.* 1993), (2) failure for piles in sand usually indicates excessive pile settlement and not an unstable condition, and (3) pile group foundations have reserve capacities which reduce the probabilities of failure by up to one order of magnitude (Tang and Gilbert 1993).

2.6 Sources of Inaccuracy in the API Method

The physical reasons which contribute to the discrepancy between measured and predicted pile capacities are discussed next.

2.6.1 Statistical Considerations

The primary inadequacy of the API method is the small size of the database on which it is based. Unfortunately, the available data is very limited and generally of poor quality. Additionally, the available data is for piles that are typically smaller in diameter, shorter in length, and have less capacity than piles used in the offshore environments. Thus, it is not possible to resolve many of the design uncertainties using data alone.

There is little or no data to support API RP-2A recommendations for the axial capacity of piles in gravely and silty soils, where the recommendations are primarily extensions of the correlations in sand. Accordingly, the degree of confidence in the recommendations falls sharply in silty and gravely soils. The general absence of such soils in the marine environment has made this problem one of limited interest.

The properties (δ, f_{lim}, N_q, q_{lim}, K) used in the analysis started with simple values that were based on a few case histories. As additional case histories became available, the properties were modified into their present form. It is probably time to modify them further to eliminate some of the scatter. Olson (1990) showed that it was possible to eliminate the depth dependency of Q_c/Q_m just by selecting different properties. The current database is larger than the one used by Olson (1990) and the best fit properties appear to be somewhat different. The properties should be looked at as reflecting current data and understanding and not as some sort of code.

2.6.2 Installation Effects

The design method uncouples the final axial capacity of a pile from the events taking place during driving. Pile driving results in changes in the state of stress, relative density, particle size, particle shape, and gradation. Additionally, pile driving densifies the soil in the vicinity of the pile but may also produce a loosened zone immediately in contact with the pile (Robinsky and Morrison 1964).

Plugging of pipe piles is another important phenomenon which is not adequately accounted for in the current recommendations. An open-ended pipe pile is said to be plugged when the soil inside the pile moves down with the pile, resulting in the pile becoming effectively closed-ended. Plugging is believed to result in an increase in horizontal stresses between the pile and the surrounding soil, which results in an increase in skin friction. Under static conditions, both theoretical analyses and experiments have shown that plugging occurs with only a small pile penetration (Murff *et al.* 1990). However, measured heights of plugs in the field are far in excess of what would be predicted from static considerations (Tsien 1986). It is likely that the observed lengths of the plug can be explained by the dynamics of the plug, as well as a build up of transient pore pressures during driving, which gives rise to cyclic mobility of the sand. Cyclic mobility results in a reduction in the ambient effective stress with an associated reduction in the frictional capacity of the core (plug). During static loading, excess pore pressures dissipate, resulting in high frictional resistances and arching within the plug.

2.6.3 Axial Capacity Mechanism

2.6.3.1 Pile Movements

Typical load-settlement curves for pipe piles in sand indicate that there is no plunging failure load, and that loads continue to increase with increasing settlement. The magnitude of the design load should be tied to the load-settlement characteristics of the pile/soil system, and the tolerance of the structure to foundation movements. Predictions of pile movements can be made using comparatively simple models based on Mindlin's equation (Poulos and Davis 1968) or the t-z approach (Coyle and Sulaiman 1967 and Aschenbrenner and Olson 1984).

Static methods of analysis usually provide equations for tip bearing capacity (Eq. 2.5) that come from bearing capacity theory. The supporting theories are usually of the rigid sliding block types which cannot apply to pile tip bearing capacity. Tip stresses may be controlled more by the compressibility of the soil below the tip than by bearing capacity theory (Vesic 1967).

2.6.3.2 State of Stress in the Vicinity of the Pile

The API recommended practice implies that the vertical effective stress close to the pile is the free field vertical effective stress. It is clear that loads transferred between the pile and the surrounding soil must perturb the previous free field stresses, but by indeterminate amounts, especially for pile groups.

Experimental evidence (Vesic 1970) indicates that the rate of increase of unit skin friction and end bearing reduces with depth, probably due to the stress dependency of K and δ. The API recommended practice adopts a simplified approach of setting upper limits on side shear and end bearing. This approach can be justified on the basis of simplicity but should be replaced by a stress dependent variation in N_q, K, and δ, *e.g.*, as recommended by Kulhawy (1984). Olson and Iskander (1994) showed that scatter between measured and predicted pile capacities could be reduced using a set of simple logarithmic functions relating blow counts to N_q, K, and δ. Further, Olson and Iskander (2009) demonstrated that the best fit parameters may not necessarily fit pre-conceived notions of what these parameters ought to be.

Load transfer between a pile and the surrounding soil is usually determined by measuring the axial load in the pile as a function of depth. Such measurements typically indicate zones of reduced side shear near both the top and bottom of piles (Vesic 1970). Reduction in load transfer near the tip is attributed to the loss of support of the soil adjacent to the pile near the tip, which results from compression of the underlying soil layers, due to mobilizing the tip's bearing capacity. Reduction in the load transfer near the top may be due to pile shaking during driving. Go (1990) suggests that neglecting the load transfer for one diameter at the tip of the pile increases the accuracy of the predicted axial capacity.

A different kind of end effect develops when the inside of an open-ended steel pipe pile is drilled out. Two case histories were presented by Olson and Al-Shafei (1988) show that the un-stressing of the pile tip results in the need for considerably larger tip settlements to develop tip capacity.

Although residual stresses do not affect the total pile capacity, they may influence the load-transfer mechanism, thereby reducing pile capacity developed at a particular settlement (Darrag 1987).

2.6.3.3 Friction Between Pile and Soil

Some engineers believe that pile-soil shearing displacement occurs at the interface between the pile and the soil, thus making the soil-pile friction angle, δ, the relevant parameter, rather than the soil friction angle, ϕ. Evidence of this view comes from pull-out tests in which the pile surface is clean. The counter view that failure occurs within the soil is supported by occasional pull out tests in which the pile is coated with sand. Studies by Uesugi *et al* (1988) suggest that failure occurs between the soil and the pile unless the pile is very rough.

There is also limited evidence that pile driving crushes the adjacent sand so the material surrounding the pile may differ significantly from the natural sand. In such a case, sand/pile shearing tests that have been used by some investigators (*e.g.* Furlow 1968) become less relevant.

The angle of internal friction of sand decreases as the stress level increases (de Beer 1967, Vesic 1967). It is therefore reasonable that δ should also decrease with stress level. Randolph (1985) advocates replacing δ with the angle of friction at large strains, ϕ_{CV}, because ϕ_{CV} is not stress dependent.

2.6.3.4 Tension versus Compression

Some authors believe that the side shear in tension should be less than the side shear in compression by perhaps 30% as reflected in the 1982 API RP-2A. Skin friction can be reduced in tension over compression due to (1) strain compatibility of the soil-pile system, and (2) reduction in the pile's diameter due to Poisson's effect. Both factors would reduce the normal stress acting at the soil-pile interface.

A finite element study, (Nystrom 1984) employing the Modified Cam Clay model in drained conditions (frictional resistance) suggested that the difference in the axial capacity of piles in tension and compression due to axial straining of the soil is less than 1%.

The existing database does not indicate that tensile capacities in side shear are less than compressive capacities; however, such a difference is reasonable and the lack of supporting data may result from data scatter or errors in computing the tip bearing capacity. It is possible that curve fitting could be performed using side shears in tension of only say 70% to 80% of those in compression, and equally good fits would be obtained, but with different parameters.

2.6.3.5 Type and Rate of Loading

Unlike lateral loading, API RP-2A does not provide guidelines for the effects of cyclic loading and rate of loading on the axial capacity of piles. There is evidence that cyclic tensile loading, in particular, may significantly reduce pile capacity (Kraft 1990).

2.6.4 Piles Driven through Clays into Sand

Tomlinson (1971) showed that driving of piles through soft clay and into stiff clays results in a significant reduction in the side shear in the stiff clay for a substantial number of pile diameters below the interface. Briaud and Anderson (1987) used API guidelines and made a similar observation for square concrete piles penetrating from clay into sand.

In this study, tests were excluded from the database if more than 15% of the calculated side shear came from clay layers. Nevertheless, evidence exists in the database that suggests that clay layers cause a significant reduction in side shear in lower sand layers, in agreement with Briaud and Anderson (1987).

2.7 Conclusions

The API Method has become the accepted industry standard for pile design as demonstrated by its widespread use. The criticisms cited herein are, to a great extent, a product of inadequate understanding of pile behavior. The criticisms apply not only to the API method but also to many other experience based deign approaches. The number of useful load tests in predominantly cohesionless soils is small, especially for steel pipe piles. The number of load tests is less than the number of degrees of freedom in the system; accordingly there are various combinations of properties that yield equivalent results. Additionally, the suspect quality of many of the available case histories contributes to the scatter between measured and predicted capacities. Thus, many of the important issues cannot be resolved on the basis of the available data alone.

Experimental research on the physical processes that control the behavior of piles during installation and loading is required in order to resolve the important uncertainties in axial capacity predictions. Considering the large number of variables involved in axial pile capacity development, as well as the high cost of load tests on full scale instrumented piles, it is unlikely that a sufficient number of full scale load tests can be performed to resolve the problem. Experimental research on model piles can however be used to resolve some of the important design uncertainties. One of the objectives of this monograph is to provide a blueprint for developing the necessary facilities required to perform such research.

References

American Petroleum Institute (API): API Recommended Practice for Planning, Designing, and Constructing Fixed Offshore Platforms, Working Stress Design, 21st edn. (RP 2A-WSD), American Petroleum Institute (2000)

Aschenbrenner, T.B., Olson, R.E.: Prediction of Settlement of Single Piles in Clay. In: Analysis and Design of Pile Foundations, pp. 41–58. ASCE Special Technical Publication (1984)

Briaud, J.L., Anderson, J.: Evaluation of API Method Using 98 vertical pile load tests. In: Proc. 19th Offshore Technology Conf. (1987)

Coyle, H.M., Sulaiman, I.H.: Skin Friction for Steel Piles in Sand. J. Soil Mech. and Found. Div., ASCE 93(SM6), 261–278 (1967)

Darrag, A.A.: Capacity of Driven Piles in Cohesionless Soils Including Residual Stresses, Ph.D. Dissertation, School of Civil Engineering, Purdue University (1987)

Davisson, M.T.: High Capacity Piles. In: Soil Mech. Div., Ill. Sect., pp. 81–112. ASCE, Chicago (1973)

de Beer, E.E.: Bearing Capacity and Settlement of Shallow Foundations on Sand. In: Proc. Bearing Capacity and Settlement of Foundations, Department of Civil Engineering, Duke University, Durham, N.C., pp. 15–33 (1967)

Dennis Jr., N.D., Olson, R.E.: Axial Capacity of Steel Pipe Piles in Clay. In: Proc. Geotechnical Practice in Offshore Engineering, ASCE, Geotechnical Engr. Div. Specialty Conf., University of Texas at Austin, pp. 370–388 (1983a)

Dennis Jr., N.D., Olson, R.E.: Axial Capacity of Steel Pipe Piles in Sand. In: Proc. Geotechnical Practice in Offshore Engineering., ASCE Geotechnical Engr. Div. Specialty Conf., Univ. of Texas at Austin, pp. 389–402 (1983b)

El-Mabsout, M.: A Finite Element Model for the Analysis of Pile Driving, Ph.D. Dissertation, University of Texas at Austin (1991)

Focht, J.A., O'Neill, M.: International State of The Practice for Design and Installation of Axially Loaded Piles. In: Proc. 11th International Conf. on Soil Mechanics and Foundation Engineering, vol. 1, pp. 187–209 (1985)

Furlow, C.R.: Pile Tests, Jonesville Lock and Dam Quachita and Black Rivers Arkansas and Louisiana, In: Technical Report S-68-10, U.S. Army Engineer Waterways Experiment Station (1968)

Go, V.: Axial Static Load Capacity of Untapered Piles in Sand, M.Sc. in Engr. Thesis, The University of Texas at Austin (1990)

Hossain, M.K., Briaud, J.L.: Pipe Piles in Sand - An Improvement for the API RP 2A. In: Proc. Offshore Technology Conf. (1992)

Iskander, M.G., Olson, R.E.: Review of API Guidelines for Pipe Piles in Sand. In: Proc. Civil Engineering in the Oceans V, pp. 798–812. ASCE (1992)

Kraft, L.M.: Computing Axial Pile Capacity in Sands for Offshore Conditions. Marine Geotechnology 9, 61–92 (1990)

Kulhawy, F.H.: Limiting Tip and Side Resistance, Fact or Fallacy? In: Meyer, J.R. (ed.) Proc. ASCE Symposium on Analysis and Design of Pile Foundation, San Francisco, CA (1984)

Lacasse, S., Goulois, A.M.: Uncertainty in API Parameters for Prediction of Axial Capacity of Driven Piles in Sand. In: Proc. 21st Offshore Technology Conf., Houston, OTC paper 6001, pp. 353–358 (1989)

Lacasse, S.: Uncertainties in Offshore Geotechnical Engineering, International Survey of API RP-2A Design Parameters for Axial Capacity of Driven Piles in Sand, Norwegian Geotechnical Institute, Report No: 85307-14 (1988)

Lings, M.L.: The Skin Friction of Driven Piles in Sand, M.Sc. Thesis, Imperial College of Science and Technology, University of London (1985)

Murff, J.D., Rains, D., Randolph, M.: Soil Plug Behavior of Piles in Sand. In: Proc. 22nd Offshore Technology Conf., Houston (1990)

Meyerhof, G.G.: Bearing Capacity and Settlement of Pile Foundations. Journal, Geotechnical Engr. Div., ASCE 102(GT3), 196–228 (1976)

Notingham, L.E.: Use of Quasi-Static Friction Cone Penetrometer Data to Predict Load Capacity of Displacement Piles, Ph.D. Dissertation, University of Florida (1975)

Nystrom, G.: Finite-Strain Axial Analysis of Piles in Clay. In: Meyer, J.R. (ed.) Proc. ASCE Symposium on Analysis and Design of Pile Foundation, San Francisco, CA (1984)

Olson, R.E., Dennis, N.D.: Review and Compilation of Pile Test Results, Axial Pile Capacity, Report to the American Petroleum Institute, Geot. Engr. Report No. GR81-14, Department of Civil Engineering, Univ. of Texas, Austin, TX (1982)

Olson, R.E., Al-Shafei, K.S.: Axial Load Capacities of Steel Pipe Piles in Sand. In: Proc. 2nd Intern. Conf. on Case Histories in Geot. Engr., Univ. of Missouri, Rolla, vol. 3, pp. 1731–1738 (1988)

Olson, R.E.: Axial Load Capacity of Steel Pipe Piles. In: Proc. 22nd Offshore Technology Conf., Paper 6419 (1990)

Olson, R.E., Iskander, M.G.: Axial Load Capacity of Pipe Piles in Sands. In: Proc. 1st Regional Conf. and Exhibition on Advanced Technology in Civil Engineering, Manama, Bahrain, pp. 383–394. ASCE–SAS (1994)

Olson, R., Iskander, M.: Axial Load Capacity of Un-Tapered Piles In Cohesionless Soils. In: Contemporary Topics in Deep Foundations, GSP No. 185, pp. 231–238. ASCE (2009)

Peck, R.B., Hansen, W.E., Thornburn, T.H.: Foundation Engineering, p. 114. John Wiley & Sons, NYC (1974)

Pelletier, J.H., Murff, J.D., Young, A.C.: Historical Development and Assessment of the Current API Design Methods for Axially Loaded Piles. In: Proc. Offshore Technology Center, Houston, TX, OTC paper 7157, pp. 253–282 (1993)

Poulos, H.G., Davis, E.H.: Pile Foundation Analysis and Design. John Wiley and Sons, NY (1968)

Randolph, M.F.: Capacities of Piles Driven into Dense Sands. Presentation to the 11th Conf. of Soil Mechanics and Foundation Engineering, San Francisco, Session 4C (1985)

Robinsky, E.I., Morrison, C.F.: Sand Displacement and Compaction Around Model Friction Piles. Canadian Geotechnical Journal 1(2), 81–93 (1964)

Sellgren, E.: Friction Piles in Non-Cohesive Soils. Evaluation from Pressuremeter Tests, Ph.D. Dissertation, Chalmers University of Technology, Goteborg, Sweden (1982)

Tang, W.H., Gilbert, R.B.: Case Study of Offshore Pile System Reliability. In: Proc. 25th Offshore Technology Conf., pp. 677–686 (1993)

Tomlinson, M.J.: Some Effects of Pile Driving on Skin Friction, Behavior of Piles. Institution of Civil Engrs., London, pp. 107–114, Response to Discussion on pp. 149–152 (1971)

Toolan, F.E., Lings, M.L., Mirza, U.A.: An Appraisal of API RP 2A Recommendations for Determining Skin Friction of Piles in Sand. In: Proc. 22nd Offshore Technology Conf., Houston, TX, Paper No: OTC 6422 (1990)

Tsein, S.I.: Shaft Frictional Resistance of Long Pipe Piles Driven into Dense Sands. In: Proc. 18th Offshore Technology Conf., Houston, TX (1986)

Uesugi, M., Kishida, H., Tsubakihara, Y.: Behavior of Sand Particles in Sand-Steel Friction. Soils and Foundations 28(1), 107–118 (1988)

Vesic, A.S.: A Study of Bearing Capacity of Deep Foundations, Final Report on Project B-189, Georgia Tech University, Atlanta (1967)

Vesic, A.S.: Load Transfer in Pile-Soil Systems. In: Fang, Dismuke (eds.) Proc., Conf. on Design and Installation of Pile Foundations and Cellular Structures, pp. 47–73. Envo Publishing Co. (1970)

Vesic, A.S.: Expansion of Cavities in Infinite Soil Mass. J. Soil Mech. Found. Div., ASCE 98(SM3), 265–290 (1972)

3. Installation Effects on the Capacity of Piles in Sand

Abstract. This chapter examines some of the most important physical processes that occur during pile driving. These processes include pile plugging, build-up and decay of pore pressures, and redistribution of stresses around the pile.

3.1 Introduction

Driving results in significant changes in the state of stress and soil fabric around piles. These changes influence the behavior of piles during installation and subsequent loading. Typically, design guidelines do not explicitly account for these changes, but often use empirical factors to calibrate for their effects. As discussed in chapter two, a better understanding of the physical mechanisms that control the behavior of piles is required in order to improve axial capacity predictions of pipe piles in sand. In this chapter, pile plugging, build-up and decay of pore pressures, and redistribution of stresses around the pile are examined.

3.2 Plugging of Piles

During installation of open ended pipe piles, the soil enters the pile at a rate which is equal to, or larger than, the rate of pile penetration. This mode of penetration is often referred to as *coring* or *cookie cutter*. As penetration progresses, the soil *core* inside the pile may develop sufficient frictional resistance along the inner pile wall to prevent further soil intrusion, causing the pile to become *plugged*[1]. Plugging is important, not only because it directly contributes to the tip bearing capacity, but also because it indirectly contributes to the developed shaft capacity. A plugged pile displaces more soil than a pile that penetrates in a coring mode, which increases the effective stresses surrounding the pile. Plugging also influences the dynamic behavior of piles, which complicates the dynamic analyses of piles (Paikowsky 1989, Paikowsky and Whitman 1990, Raines *et al.* 1992).

[1] Technically, the soil *core* should be referred to as a *plug* only when it is wedged against the pile, thus preventing any additional soil entry into the pile. Unfortunately, the term plug has often been used to refer to the core regardless of its state during installation (Paikowsky 1990).

M. Iskander: Behavior of Pipe Piles in Sand, SSGG, pp. 25–33.
springerlink.com

3.2.1 *Plugging of Piles during Installation*

The common view is that most piles that plug during static loading do not plug during driving (Paik and Lee 1993). Smith and Chow (1982) attributed this phenomenon to a combination of inertial effects and an increase in the dynamic bearing capacity factor, *Nq*, over its static value. Additionally, Smith *et al.* (1986) argued that the driven pile mobilizes all of its internal and external friction intermittently during penetration and, as a result, the soil core advances up the pile[2].

Conversely, Paikowsky *et al.* (1989) have argued that during driving "*the pile plugging phenomenon is of frequent occurrence and is of greater significance than that presently accorded it by the profession.*" The degree of plugging depends on the hammer characteristics, soil properties, and the pile's frictional resistance (Brucy *et al.* 1991).

In rare occasions, such as driving in very dense sands, piles may plug and impede driving. Plugging may result in a situation where the available pile hammer cannot drive the pile to the design depth. The problem is more critical for piles with thickened walls near the mudline (surface), such as the piles used to resist lateral loading (Murff 1990). If the pile plugs and "refuses" prematurely, the required thick section may end up above the mudline. If a plug forms during the installation of an offshore pile, the plug is typically removed by drilling or jetting[3]. The effects of the removal of the soil plug on the final pile capacity are controversial. Brucy *et al.* (1991) claimed that static loading results are not changed by partial removal of the soil plug. Other studies indicate that jetting results in a substantial reduction in the over-all pile capacity, even if the pile is re-driven (McClelland and Focht 1955).

3.2.2 *Plugging of Piles during Static Loading*

It is generally accepted that the soil core develops sufficient frictional resistance during static loading to mobilize a plug (Paik and Lee 1993). The current design practice is to limit the end bearing capacity of a pile to the smaller of (1) the value of the frictional capacity of the soil core, or (2) the end bearing capacity computed from bearing capacity theory. The increase in the shaft's skin friction due to plugging is not accounted for. Field observations suggest that the current design practice does not adequately account for plugging in a number of ways. First, the

---

2 Smith *et al.* (1986) also concluded that "*in general, analysis predicts that closed-ended piles will be easier to drive than open ended piles.*" This conclusion is contrary to the experience of many engineers, which raises doubt about the validity of their numerical approach.

3 Typically, construction documents specify a design depth for piles, which cannot be changed if the pile meets early refusal. The high daily cost of offshore construction, with barge time costing up $500,000/day, typically forces construction personnel into jetting the plug out instead of securing a larger hammer.

observed final heights of the soil cores is larger than those predicted using the design method (Tsien 1986). Second, the formation of the plug appears to be erratic, and not follow the design model (Paikowsky *et al.* 1989). Third, the settlement of open-ended plugged piles appears to be larger than that of closed ended piles (Kishida 1967, Randolph 1985, Briaud and Audibert 1990).

Both analytical and experimental studies have concluded that plugging results from arching of the soil in the first few diameters (Leong and Randolph 1991 and Kishida and Isemoto 1977). The available soil friction along the inside of pipe piles is significantly higher than common design values for the external skin friction, for sands (Murff *et al.* 1990), clays (Randolph 1987), and even for calcareous soils (Randolph *et al.* 1990). Small deformations on the order of 0.1 in. are required to mobilize arching, particularly in dense sands (Kishida *et al.* 1985).

Piles which plug during static loading may, nevertheless, have smaller tip bearing capacities than their closed-end counterparts (Szechy 1958). O'Neill and Raines (1991) concluded that the reduction in tip bearing capacity is primarily due to the plug's compressibility.

3.2.3 Plugging of Piles during Cyclic Loading

Offshore oil and gas platforms are subjected to substantial forces due to wave loading, which are therefore cyclic in nature. Wave loading can generate pore pressures in at least one of two ways (Lee and Poulos 1988). First, in the near-shore region, where water depths are shallow, the traveling pressure wave can induce cyclic shear stresses within the soil mass which may cause a progressive build-up of excess pore pressures over a wide area. Second, the repeated cyclic loading of the pile foundations may result in a buildup of pore pressures locally around the pile shaft.

Arching of soils inside the pile depends on the ability of the soil to drain pore pressures and develop high frictional stresses along the soil/pile interface. Under cyclic loading, the soil core may become partially drained, which may prevent pile plugging (Randolph *et al.* 1991).

3.3 Buildup of Pore Pressures during Driving

3.3.1 Excess Pore Pressures in Clays

For piles driven in clay, it is generally accepted that driving results in an increase in both the total stress and pore water pressure over their static values. Excess pore pressures up to eight times the free field effective stresses, and at distances as far as 15 diameters away from the pile have been measured (*e.g.* Reese and Seed 1955, Lo and Stermac 1965, D'Appolonia and Lambe 1971, Robertson *et al.* 1990). Excess pore pressures typically result in easier driving and in a time dependent increase in pile capacity, which is referred to as *pile setup* (Vesic 1975).

Excess pore pressures in clays result from a combination of cyclic strain and an increase in the total stress, which take place during pile driving.

3.3.2 Excess Pore Pressures in Sands

The soils surrounding piles driven in sand are subject to cyclic strain and an increase in total stress, like piles driven in clay. The generation of excess pore pressure due to cyclic loading in sand has long been widely recognized (Seed and Lee 1966). Nevertheless, buildup of pore pressures due to pile driving is typically not recognized. A few investigators, including York *et al* (1995), Chellis (1961) and Datta (1983), did, however, suggest the possibility of pile setup in sands.

Both field and laboratory studies have provided evidence that pile driving may result in a buildup of pore water pressure in sands (Table 3.1). Plantema (1948) measured excess pore pressure ratio (u/σ_v) ranging between 0.0–0.2 for production-size piles. The pore pressures dissipated in 5–45 minutes after driving, for coarse and fine sand, respectively. Moller and Bergdhal (1981) measured dynamic pore pressures during driving of laboratory model piles. Dynamic pore pressures dissipated completely after each blow and there was no overall rise in the static pore pressures. Buildup of pore pressures is more commonly observed when sands are inter-bedded with clays (*e.g.* Lacy 1981 and Moe *et al.* 1981).

Table 3.1 Pore Pressures Induced Due To Pile Driving in Sand

Reference	Soil Type	Pore Pressure Ratio
Plantema 1948	Coarse Sand	0.0 —0.1
Plantema 1948	Silty Fine Sand	0.0 — 0.2
Moller and Bergdhal 1981	Fine Sand	-0.2 — 0.15

3.3.3 *Effects of Pile Setup on Drivability*

The role of excess pore pressures in reducing the energy required to drive piles in clay is well documented (Reese 1981). When pore pressures are artificially raised, the end bearing and skin friction are reduced (Armishaw and Cox 1979). Buildup of transient excess pore pressure in the sand during driving is therefore likely to reduce the driving resistance. Because sands have a high permeability, buildup of pore pressures is likely to have small, transient effects on skin friction. The greatest effect of a buildup of pore pressures is possibly a reduction in the frictional stresses within the soil core in the first few pile diameters from the tip. The stipulated reduction in frictional stresses may preclude the formation of a plug as discussed by Randolph *et al.* (1991) for cyclic loading.

Pile driving results in densification of all sands immediately below the pile tip, regardless of their initial relative density (Szechy 1961). This densification extends within the first few diameters of the soil core. Densification is an important ingredient in the formation of an arch and promoting plugging. In its dense state, the sand below the tip and within the core is not capable of liquefying under cyclic loading. Nevertheless, the sand is capable of reaching a state which is referred to as cyclic mobility. When sands become cyclically mobile, their pore pressures increase, which leads to a decrease in their effective stresses, modulus, and frictional resistance.

Pile plugging is typically reported in very dense sands. These sands are typically dilative and, therefore, have a reduced ability to generate excess pore pressures during cyclic loading and/or pile driving.

In Chapter 9, measurement of a transient buildup of pore pressures during driving is presented. This build-up is likely sufficient to destroy the arching mechanisms, which helps the soil core advance up the pile. In rare occasions, such as driving in very dense sands, build-up of pore water pressure is not sufficiently large to destroy soil arching, and a plug forms.

3.4 Stress Redistribution Due to Driving

Pile driving results in changes in the state of stress and soil fabric surrounding the pile. These changes are typically incorporated in various design approaches using empirical earth pressure coefficients and/or limits on the stresses that can be mobilized in skin friction. An alternate way of introducing a limit on skin friction is the concept of the critical depth, which was first introduced by Vesic (1967), and later adopted by others (*e.g.* Meyerhof 1976).

Kulhawy (1984) expressed the view that the perception of a critical depth results from the depth dependency of the friction angle, δ, and the earth pressure coefficient, K. Kulhawy's view is supported by observations that the skin friction increases with depth at a progressively decreasing rate (*e.g.* Vesic 1967 and Coyle and Castello 1981).

Kraft (1991) argued that the critical depth concept is a result of arching of the sand around the pile. Kraft's view is supported by Robinsky and Morisons's (1964) study which concluded that as their pile was pushed into the ground, it densified the soil below the tip. However, as the pile penetrated deeper it dragged the soil in its immediate vicinity thus creating a sleeve of loose sand and setting up ideal conditions for arching. Robinsky *et al.* (1964) argued that "*a cylinder of dense sand, originally compacted by the pile tip, encircles the loosened sands and prevents, by arching, the development of full lateral earth pressure on the pile.*" Szechy (1961) also measured increases in void ratio of approximately 5% adjacent to the lower half of driven piles.

Lehane *et al.* (1993) and Lehane and Jardine (1994) indicate that the earth pressure coefficient is a function of relative density. Both Olson (1990) and Kraft (1990) provide supporting evidence by introducing improved methods for axial capacity predictions which relate the skin friction to relative density. Arching may depend on relative density, which lends support to the pioneering work of Robinsky and his coworkers.

The controversy regarding the critical depth results from instrumented load tests which show conflicting skin friction distributions. Khan *et al.* (1992) concluded that "*virtually all of the shapes that have been advocated not only can occur but could all occur at various settlements in a single pile loading case.*"

3.5 Conclusions

The views of practitioners and researchers regarding the effects of the installation process on the capacity of piles are summarized in this chapter. Many of the views expressed are controversial, which highlights the complexity of the axial capacity mechanism and the scarcity of high-quality instrumented load test data. At this stage, the relationship between the various mechanisms that occur during pile driving is somewhat speculative.

This monograph presents a detailed blue print for developing experimental facilities necessary to identify these processes. These facilities include a unique instrumented double-walled pipe-pile that is used to delineate the frictional stresses acting against the external and internal surfaces of the pile. The pile is fitted with miniature pore pressure transducers to monitor the generation of pore water pressure during installation and loading. A fast automatic laboratory pile hammer capable of representing the phenomena that occur during pile driving was also developed and used. Finally, a pressure chamber; feedback control system; data acquisition system; loading frame; sand handling, pluviating, saturating, and drying apparatus have been integrated to allow convenient load testing of piles under simulated field conditions. The experimental apparatus is presented with sufficient details to allow readers to duplicate or modify the design to suit their own research needs.

References

Armishaw, J.W., Cox, D.W.: The Effects of Changes in Pore Pressures on The Carrying Capacities and Settlements of Driven Piles End Bearing in a Sand and Gravel Stratum, Recent Developments in The Design and Construction of Piles, Institute of Civil Engineers, London, pp. 227–236 (1979)

Briaud, J.L., Audibert, J.M.E.: Some Thoughts on API RP2A for Vertically Loaded Piles. In: Proc. 22nd Offshore Technology Conf., pp. 9–16 (1990)

Brucy, F., Meunier, J., Nauroy, J.F.: Behavior of Pile Plug in Sandy Soils During and After Driving. In: Proc. 23rd Offshore Technology Conf., Houston, TX, pp. 145–154 (1991)

Chellis, R.D.: Pile Foundations, 2nd edn. McGraw Hill Book Company (1961)

Coyle, H.M., Castelo, R.R.: New Design Correlations for Piles in Sand. Journal of Soil Mechanics and Foundation Division, ASCE 107, 956–986 (1981)

Datta, M.: Pore Pressure Development During Pile Driving and Its Influence on Driving Resistance. In: Proc. Geotechnical Practice in Offshore Engineering, Austin, Texas. ASCE (1983)

D'Appolonia, D.J., Lambe, T.W.: Performance of Four Foundations on End Bearing Piles. Journal of Geotechnical Engineering 8, 434–455 (1971)

Khan, N.I., Templeton, J.S., O'Neill, M.W.: Skin Friction Distribution on Piles in Sand. In: Proc. Civil Engineering in The Oceans V, pp. 783–797. ASCE (1992)

Kishida, H.: The Ultimate Bearing Capacity of Pipe Piles in Sand. In: Proc. Third Asian Regional Conf. of Soil Mechanics and Foundation Engineering, vol. 1, pp. 196–199 (1967)

Kishida, H., Isemoto, N.: Behavior of Sand Plugs in Open Ended Steel Pipe Piles. In: Proc. 9th Int. Conf. on Soil Mechanics and Foundation Engineering, Tokyo, pp. 601–604 (1977)

Kishida, H., Uesugi, M., Susumu, M.: Behavior of Dry Sands in Steel Pipe Piles. In: Proc. 8th Southeast Asian Geotechnical Conf., Kuala Lumpur (1985)

Kraft, L.: Computing Axial Pile Capacity in Sands for Offshore Conditions. Marine Geotechnology, 61–92 (1990)

Kraft, L.: Performance of Axially Loaded Pipe Piles in Sand. Journal of Geotechnical Engineering, 272–296 (1991)

Kulhawy, F.H.: Limiting Tip and Side Resistance: Fact or Fallacy? Analysis and Design of Pile Foundations, ASCE, 80–89 (1984)

Lacy, H.S.: Pile Integrity and Capacity Determined by Redriving. In: Proc. 10th Int. Conf. on Soil Mechanics and Foundation Engineering, Stockholm, vol. 2, pp. 781–786 (1981)

Lee, C.Y., Poulos, H.G.: Influence of Excess Pore Pressures on Axial Offshore Pile Response, Research Report No: R577, The University of Sydney (1988)

Lehane, B.M., Jardine, R.J., Bond, A.J., Frank, R.: Mechanisms of Shaft Friction From Instrumented Piles. Journal of Geotechnical Engineering, 19–35 (1993)

Lehane, B.M., Jardine, R.J.: Shaft Capacity of Driven Piles in Sand: A New Design Approach. In: Proc. 7th Int. Conf. on The Behavior of Offshore Structures, BOSS 1994, pp. 23–35 (1994)
Leong, E.C., Randolph, M.F.: Finite Element Analysis of Soil Plug Response. Int. Journal of Numerical and Analytical Methods in Geomechanics 15, 121–141 (1991)
Lo, K.Y., Stermac, A.G.: Induced Pore Pressures During Pile Driving Operations. In: Proc. 6th Int. Conf. on Soil Mechanics and Foundation Engineering, vol. 2, pp. 285–289 (1965)
Meyerhof, G.G.: Bearing Capacity and Settlement of Foundations. Journal of Geotechnical Engineering, ASCE 102, 197–228 (1976)
McClelland, B., Focht, J.: Padre Island Load Tests, Project Report, McClelland Engineers (Fugro-McClelland Marine Geosciences), Houston, Texas (1955)
Moe, D., Arveson, H., Holm, O.S.: Friction Bearing Pipe Piles at Calabar Port. In: Proc. 10th Int. Conf. on Soil Mechanics and Foundation Engineering, Stockholm (1981)
Moller, B., Bergdahl, U.: Dynamic Pore Pressures During Pile Driving in Fine Sand. In: Proc. Int. Conf. on Soil Mechanics and Foundation Engineering, vol. 2, pp. 791–794 (1981)
Murff, J.D.: Personal Communications. Exxon Production Research Company, Houston, TX (1990)
Murff, J.D., Raines, R.D., Randolph, M.F.: Soil Plug Behavior of Piles in Sand. In: Proc. 22nd Offshore Technology Conf., Houston, TX, pp. 25–32 (1990)
Olson, R.E.: Axial Load Capacity of Steel Pipe Piles in Sand. In: Proc. 22nd Offshore Technology Conf., Houston, TX, pp. 17–24 (1990)
O'Neill, M.W., Raines, R.D.: Load Transfer for Pipe Piles in Highly Pressured Dense Sand. Journal of Geotechnical Engineering, 1208–1226 (1991)
Paik, K.H., Lee, D.R.: Behavior of Soil Plugs in Open Ended Model Piles Driven Into Sands. Marine Geotechnology 11, 353–373 (1993)
Paikowsky, S.G.: A Static Evaluation of Soil Plug Behavior with Application To The Pile Plugging Problem, D.Sc. Thesis, MIT (1989)
Paikowsky, S.G., Whitman, R.V., Baligh, M.: A New Look at The Phenomena of Offshore Pile Plugging. Marine Geotechnology 8, 213–230 (1989)
Paikowsky, S.G.: The Mechanism of Plugging in Sand. In: Proc. 22nd Offshore Technology Conf., pp. 593–604 (1990)
Paikowsky, S.G., Whitman, R.V.: The effect of Plugging on Pile Performance and Design. Canadian Geotechnical Journal, 429–440 (1990)
Platema, G.: The occurrence of Hydrodynamic Stresses in The Pore Water of Sand Layer During Driving of Piles. In: Proc. 2nd Int. Conf. on Soil Mechanics and Foundation Engineering, Rotterdam, vol. 4, pp. 127–128 (1948)
Raines, R.D., Ugaz, O., O'Neill, M.: Driving Characteristics of Open Towe Piles in Sand. Journal of Geotechnical Engineering, 72–88 (1992)
Randolph, M.F.: Capacity of Piles Driven Into Dense Sand. Presentation to the 11th Int. Conf. of Soil Mechanics and Foundation Engineering, San Francisco (1985)
Randolph, M.F.: Modeling of The Soil Plug Response During Pile Driving. In: Proc. 9th Asian Geotechnical Conf., Bangkok, Thailand, pp. 6-1–6-14 (1987)
Randolph, M.F., Leong, E.C., Hyden, A.M., Murff, J.D.: Soil Plug Response in Open Ended Pipe Piles. Journal of Geotechnical Engineering, 743–759 (1990)
Randolph, M.F., Leong, E.C., Houlsby, G.T.: One Dimensional Analysis of Soil Plugs in Pipe Piles. Geotechnique 41, 587–598 (1991)

Reese, L.C., Seed, H.B.: Pressure Distribution Along Friction Piles. In: Proc. ASTM, vol. 55 (1955)

Reese, L.C.: Interaction of Deep Foundations with the Supporting Soils. In: Proc. 1st Indian Conf. in Ocean Engineering, Madrass, India, vol. 2, pp. 143–154 (1981)

Robertson, P.K., Woeller, D.J., Gillespie, D.: Evaluation of Excess Pore Pressures and Drainage Conditions Around Driven Piles Using The Cone Penetration Test With Pore Pressure Measurements. Canadian Geotechnical Journal, 249–254 (1990)

Robinsky, E.I., Morrison, C.F.: Sand Displacement and Compaction Around Model Friction Piles. Canadian Geotechnical Journal 1(2), 81–91 (1964)

Robinsky, E.I., Sagar, W.L., Morrison, C.F.: Effect of Shape and Volume on the Capacity of Model piles in Sand. Canadian Geotechnical Journal 1(4), 189–204 (1964)

Smith, I.M., Chow, Y.K.: Three Dimensional Analysis of Pile Drivability. In: Proc. 2nd Int. Conf. on Numerical Methods in Offshore Piling, Austin, Texas, pp. 1–19 (1982)

Smith, I.M., To, P., Wilson, S.M.: Plugging of Pipe Piles. In: Proc. 3rd Int. Conf. on Numerical Methods in Offshore Piling, Nantes, France, pp. 54–73 (1986)

Seed, H.B., Lee, K.L.: Liquefaction of Saturated Sands During Cyclic Loading. Journal of the Soil Mechanics and Foundations Division, ASCE 92, 105–134 (1966)

Szechy, C.: Tests with Tubular Piles. Acta Technica, Hungarian Academy of Science 24 (1958)

Szechy, C.: The Effects of Vibration and Driving Upon the Voids in Granular Soil Surrounding a Pile. In: Proc. 5th Int. Conf. of Soil Mechanics and Foundation Engineering, vol. 2, pp. 161–164 (1961)

Tsien, S.I.: Shaft Frictional Resistance of Long Pipe Piles Driven Into Dense Sands. In: Proceedings 18th Offshore Technology Conf., pp. 601–610 (1986)

Vesic, A.: A Study of Bearing Capacity of Deep Foundations, Final Report, Project B-189, Georgia Institute of Technology (1967)

Vesic, A.: Principles of Pile Foundations Design. Soil Mechanics Series, vol. 38. School of Civil Engineering, Duke University (1975)

York, D., Brusey, W., Clemente, F., Law, S.: Setup and Relaxation in Glacial Sand. ASCE J. Geotechnical Engr. 120(9), 1496–1513 (1995)

4. Experimental Facilities to Study the Behavior of Piles

Abstract. This chapter documents the design, fabrication, and performance of the experimental facility designed to study the behavior of piles in sand. The equipment consists of a pressure chamber; feedback control system; data acquisition system; loading frame; sand handling, pluviating, saturating, and drying apparatus have been integrated to allow convenient load testing of piles under simulated field conditions. The experimental apparatus is presented with sufficient details to allow readers to duplicate or modify the design to suit their own needs.

4.1 Introduction

One of the major obstacles to an improved prediction method for the capacity of piles in sand has been a lack of an understanding of the physical processes which control the behavior of piles during installation and loading. Ideally, such understanding should be achieved through load tests on instrumented full-scale piles. However, the costs associated with conducting many full scale tests is prohibitive, particularly for large diameter offshore piles. As a result, laboratory load tests on model piles are needed to identify the important mechanisms that influence the behavior of piles. Laboratory load tests have a few advantages over full scale tests. For example, multiple load tests under identical controlled conditions can easily be performed. It is also possible to study the various parameters affecting pile capacity by altering each parameter while maintaining the rest constant.

One of the objectives of this project has been the development of an experimental facility for performing load tests on piles under controlled laboratory conditions. This chapter documents the design, fabrication, and performance of the experimental facility designed and built to accomplish this task. Piles are installed and load tested in a laboratory pressure chamber. A loading system capable of applying loads up to 30 tons and a data acquisition system capable of sampling up to 32 channels in real time were also developed. Two additional pieces of the developed equipment are not included in this chapter; the double wall pile, and the pile hammer. The design and performance of these two instruments is presented in chapters 5 and 6 respectively.

The geotechnical centrifuge is often used to study the behavior of model piles. The centrifuge has the advantage of being able to model the gradual increase in the free field earth stresses. However, the centrifuge suffers from two disadvantages. The payload of the centrifuge is limited, which limits the size of the piles

M. Iskander: Behavior of Pipe Piles in Sand, SSGG, pp. 35–73.
springerlink.com

that can be tested. Additionally, the electrical signals of the data acquisition sensors are usually transferred to the data acquisition computer through slip rings (e.g. Schofield, 1980). The need for slip rings limits the number of data acquisition channels which can be recorded.

4.2 General Layout

The general layout of the test facilities is shown in Fig. 4.1–4.3. The facility consist of three main systems; the pressure chamber, the loading system, and the data acquisition system. Each of the three systems consists of a number of components as follows:

- *The pressure chamber system* consists of the chamber, its pressure control panel, water deairing system, and sand handling, drying, and pluviating apparatus.
- *The loading system* consists of a loading frame and an MTS feedback control system. The MTS system consists of a hydraulic actuator, high pressure pump, solenoid valve, and a control unit.
- *The data acquisition system* consists of a computer, data acquisition software and hardware, signal conditioning equipment, and a number of sensors.

4.3 The Pressure Chamber

Piles are installed and load tested in a pressure chamber[1] that is filled with sand using a method termed dry pluviation. The pressure chamber is located on the loading frame, centered below the hydraulic actuator (Fig. 4.2, 4.3). During load tests, the pressure chamber is bolted to the lower beam of the load frame using four tie rods that extend to the top of the chamber. The tie rods prevent the chamber from lifting-off the loading frame during pull-out tests, if the pile's tensile capacity exceeds the weight of the chamber (approximately 2 tons).

[1] Also known as a calibration chamber, particularly when used in conjunction with the Cone Penetration Test (CPT).

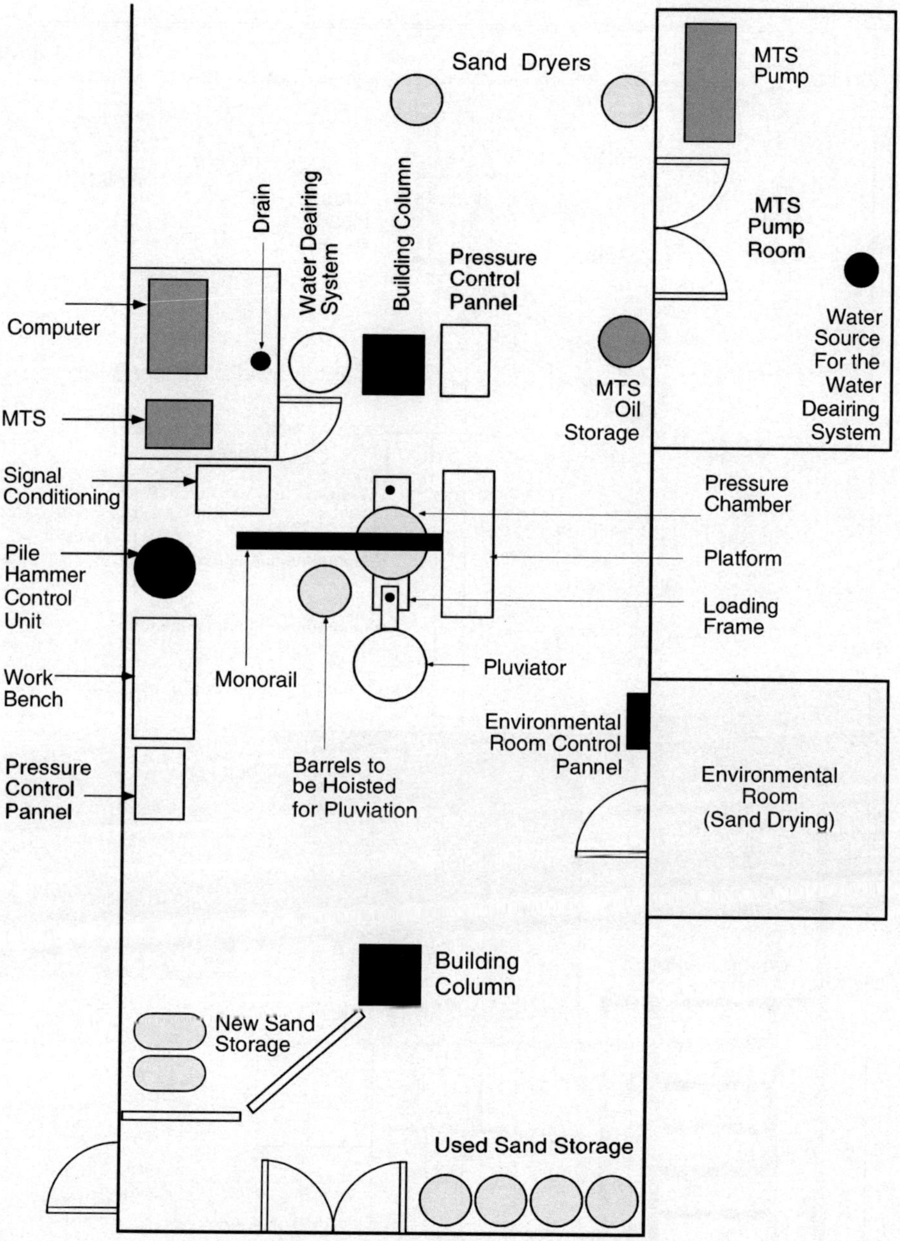

Fig. 4.1 General Layout of the Experimental Facilities

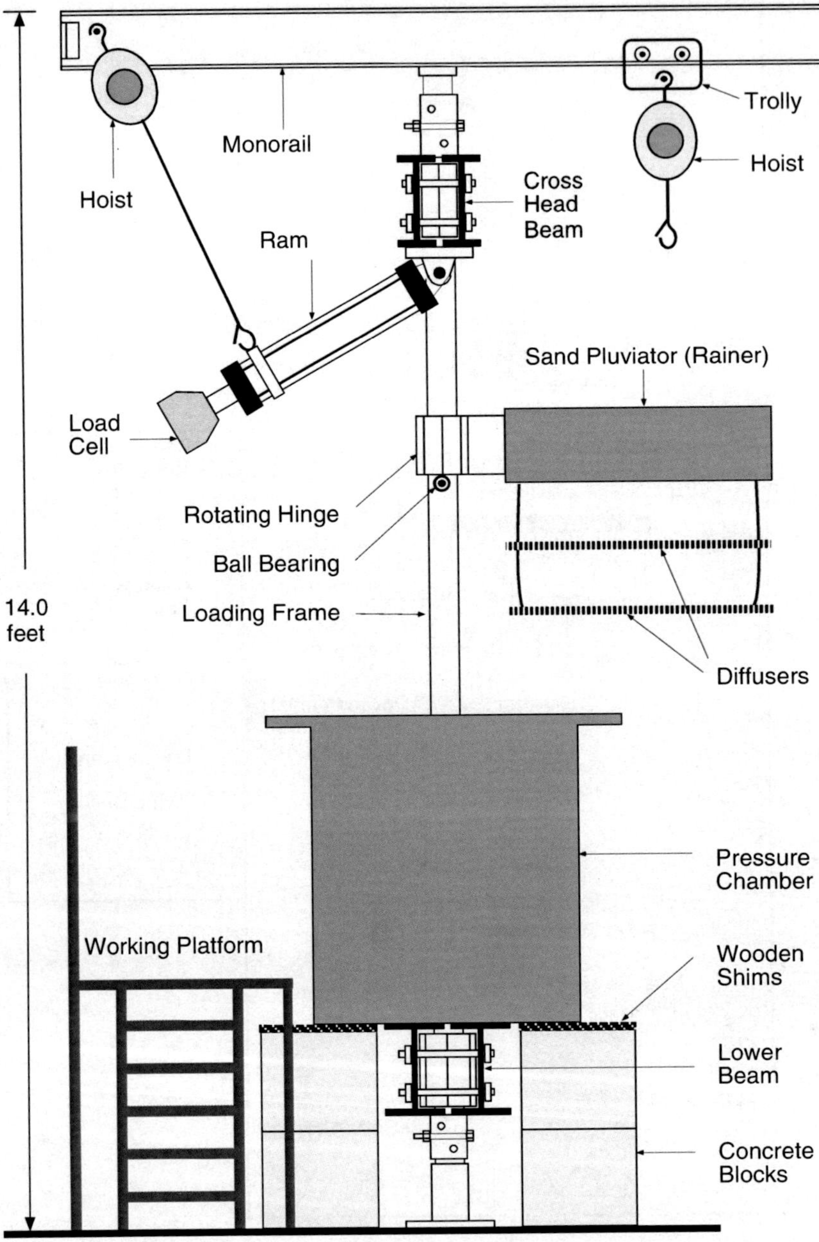

Fig. 4.2 Schematic of the Test Facility

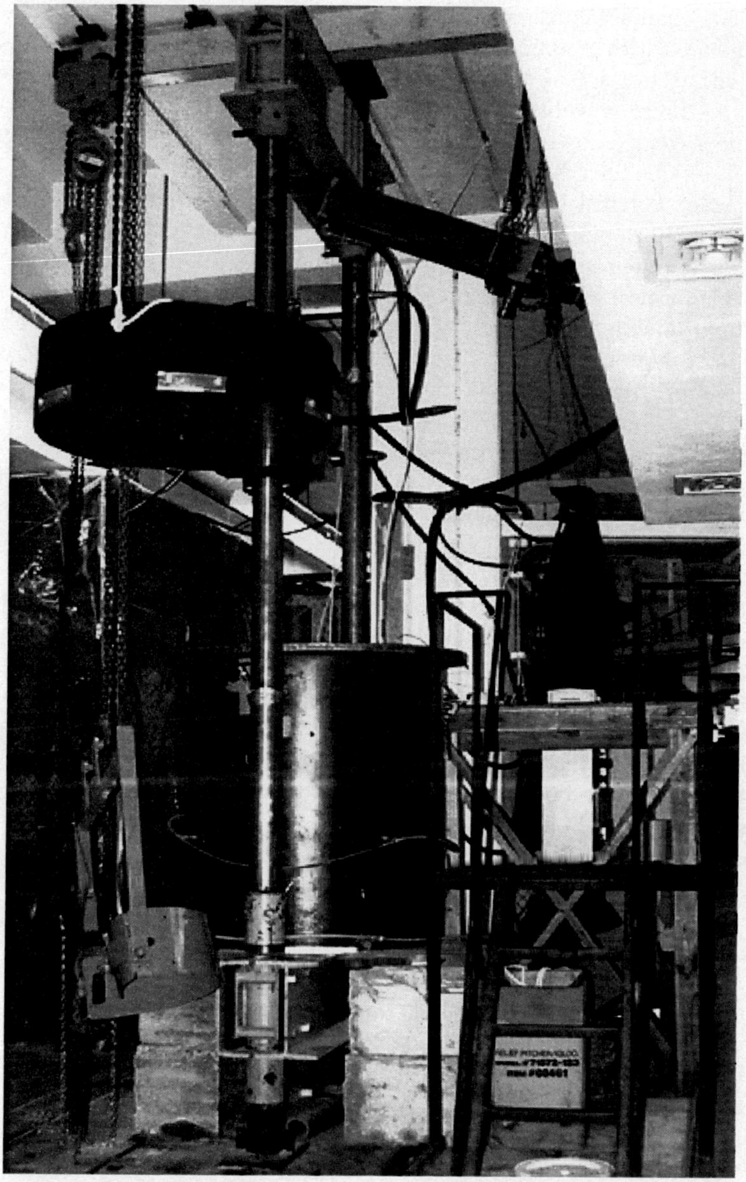

Fig. 4.3 Photograph of the Test Facility

4.3.1 Design of the Pressure Chamber

The pressure chamber consists of a cylindrical steel drum 42 in. high and 36 in. in outer diameter, with a 0.625-in. wall thickness (Fig. 4.4). The bottom of the chamber consists of a 1 in. thick steel plate which is welded to the sides. A steel flange, 1 in. thick, is welded at the top of the drum and extends 3 in. outwards. The chamber cover consists of a 1-in.-thick plate that is bolted to the flange with twelve 5/8-in.-diameter high strength bolts. Piles are installed through a 10-in.-diameter hole, which is cut at the center of the chamber cover. The hole is reinforced with a 10 in.-diameter, 3 in.-thick pipe section. The chamber is designed to withstand a 40 psi pressure, according to the American Society of Mechanical Engineers code for boilers and pressure vessels (Fouse, 1984).

A 1/16-in. thick butyl rubber sheet is bonded to the inside wall of the chamber at the top and bottom. The sheet forms a cylindrical membrane which is used to apply lateral pressure to the soil sample by injecting air between the membrane and the wall of the chamber. Two compression rings are used at the bottom and top edges of the membrane to provide a better seal. Approximately 1000 rubber squares are glued to the inner wall of the chamber, beneath the membrane, to provide a grid of air passages, which distribute air pressure behind the entire surface of the membrane. Air pressure is applied through four connections which are uniformly spaced around the drum, at a height 15 in. above its base.

Vertical pressure is applied to the sand surface using two concentric rubber tire tubes, and a cylindrical pressure ring (Fig. 4.5). The tubes are confined between the chamber, its cover plate, and the pressure ring. The pressure ring is designed to apply a contact pressure to the sand surface which matches the pressure applied by the rubber tubes as indicated in Fig. 4.6[2]. The pressure ring consists of a pipe section 10 in. in inner diameter and 7 in. high. A 10 in.-diameter, 3/8 in.-thick plate is welded to the lower end of the pipe section. Four cuts are made in the pipe to prevent transfer of any circumferential forces. A 4.3 in.-diameter hole is cut in the middle of the pressure ring to allow passage of the pile.

The sand in the chamber is saturated with deaired water which is supplied to it through four uniformly spaced water inlets near the bottom of the chamber. A 1-in. thick layer of pea gravel is placed in the bottom of the chamber to distribute the water uniformly. The gravel is covered by a geocomposite filter which prevents sand from clogging the water supply system. A second geocomposite filter is used on top of the sand to promote vertical water flow and reduce exit gradients.

The water inlets are connected to a supply tube that has a valve on each end. One valve connects the supply tube to the water deairing system, and the other connects to a drain. At the end of an experiment, water is drained out of the tank through the supply tube into the drain.

---

[2] Theoretically, the pressure ring is designed to apply an average contact stress equal to the pressure applied by the tire tubes by (1) assuming strain compatibility between the soil and the ring's lower plate, (2) assuming a modulus of subgrade reaction for the soil, and (3) determining the deformation of the ring's lower plate due to an edge moment. In practice, however, the contact stress between the ring and the soil is made uncertain by the fact that the chamber cover may apply a force to the pressure ring.

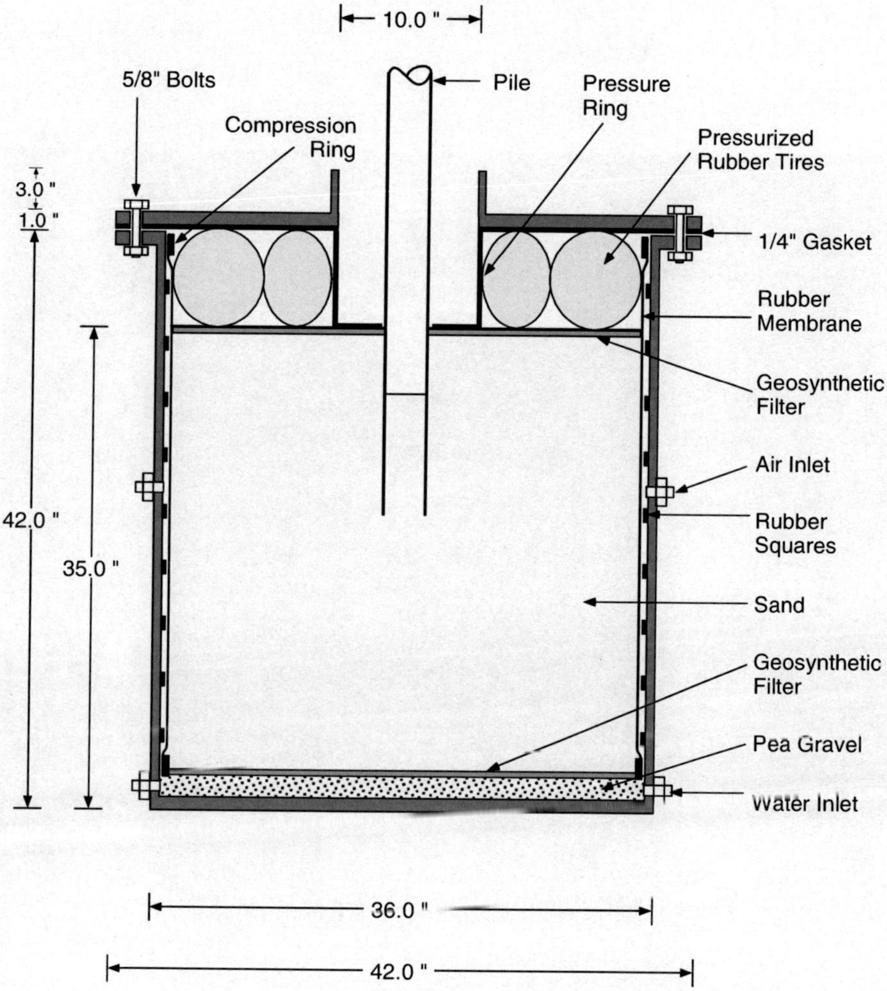

Fig. 4.4 Schematic of the Pressure Chamber

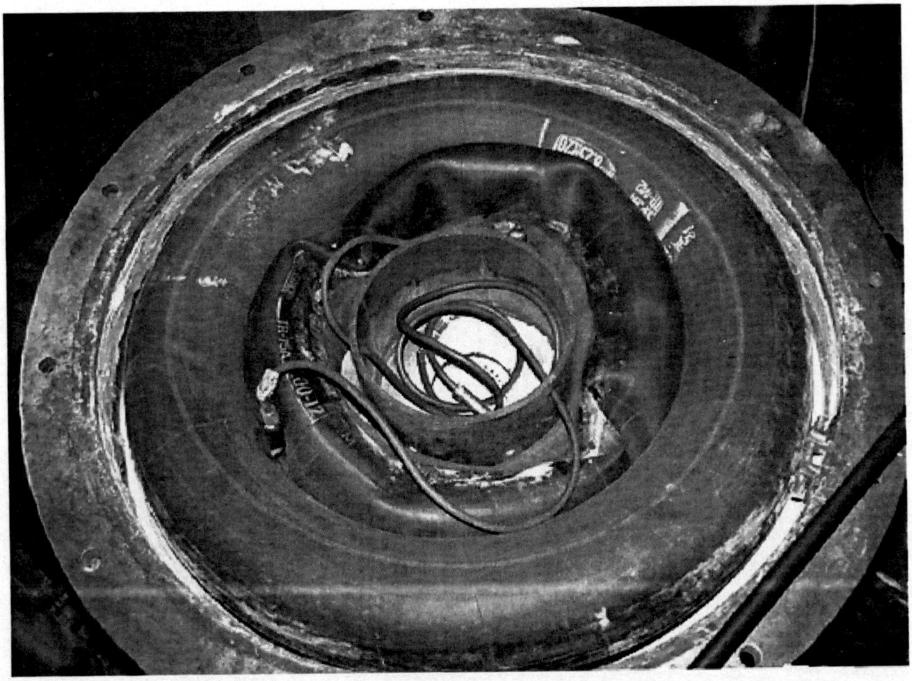

Fig. 4.5 Photograph of the Pressure Ring and Rubber Tires

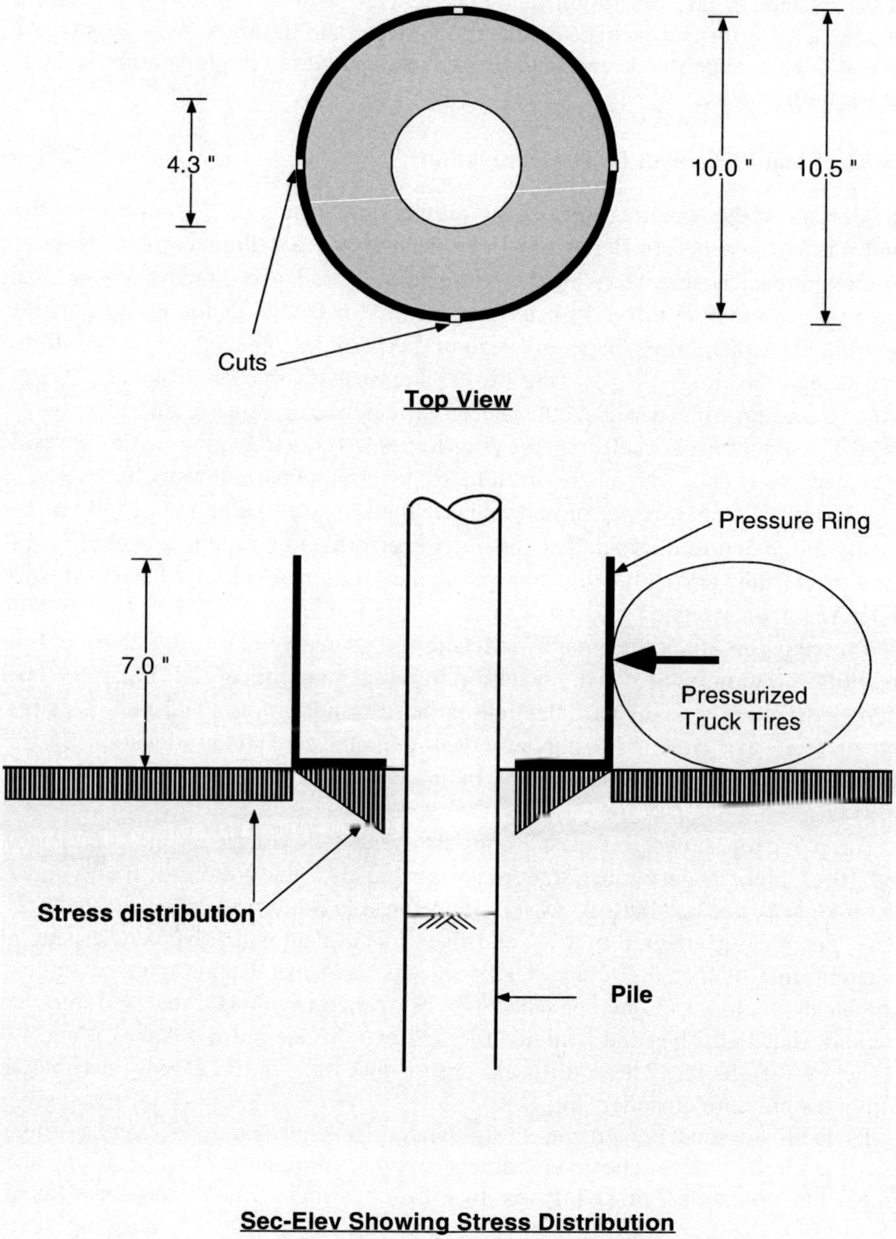

Fig. 4.6 Schematic of the Pressure Ring

4.3.2 Preparation of Soil Specimens

Preparation of a soil specimen consists of pluviating (raining) dry sand into the pressure chamber and saturation of the sand with deaired water. Dry pluviation ensures that the formed soil specimens are uniform and have a known density. At the end of each experiment, the sand is excavated and dried in preparation for pluviating it once more.

4.3.2.1 Sand Placement by Dry Pluviation

The density of the specimen formed by pluviation depends on the velocity of the sand particle just before deposition. The velocity of a falling particle depends on gravitational acceleration and air drag forces (Rad and Tumay, 1987). The magnitude of the drag force depends on a number of factors including the particle velocity, air temperature, shape, and size of the particles. The velocity of a falling particle and the drag forces acting on it increase with the fall distance. There comes a stage during particle fall where the increase in velocity due to gravitational acceleration is canceled by the drag forces. At this stage, the particles travel at a constant velocity which is referred to as the terminal particle velocity.

A battery of tests was performed to study the effect of the height of fall on the density of the deposited sand. The tests were performed by raining air-dried Oklahoma sand from varying heights into a container of known volume. The filled container was then weighed and the density of the pluviated sand determined. The sand[3] used in this study presumably reached its terminal velocity after 25 in. of fall and little variation in the density occurred for greater heights of fall (Fig. 4.7). The average dry density of the particles falling more than 25 in. is 111.5 pcf. As a result of these experiments, a pluviator was mounted at a fixed elevation 88 in. above the floor level, which is 29 in. above the top of the sand specimen (Fig. 4.2).

The pluviator shown in Fig. 4.8 consists of a steel cylinder 34 in. in diameter and 10 in. high. A perforated steel plate is welded to the bottom of the cylinder and the top is not covered. A 3/8-in. diameter hole pattern is drilled in the steel plate. The plate is covered by a second sheet metal plate which has a hole pattern matching that of the steel plate. The steel and sheet metal plates form a shutter which can be closed to prevent sand flow, or opened to pluviate the sand into the pressure chamber. The sand is rained through two diffusers which hang 10 and 17 in. below the shutter. These diffusers ensure that the sand is evenly distributed within the pressure chamber (Fig. 4.9).

To facilitate sand placement, the pluviator is connected to a circular hinge which encircles one of the load frame's circular columns (Fig. 4.8). Using the hinge, the pluviator can swing, on two ball bearings, around the load frame

[3] Very fine, uniformly graded Oklahoma sand (Simpson Formation) was used. See chapter 7 for sand properties.

column so that it is placed above the tank during pluviation, and away from the tank during load tests.

Sand placement requires lifting approximately a ton of sand to the pluviator which is located 8 ft above the floor level. To reduce the magnitude of manual labor associated with handling the sand, a hoist is used to lift the sand in 55 gallon drums. The drums are clamped in a drum frame (Moorse Manufacturing[4] model No: 1850-HD), and the frame is hoisted to the desired level. The frame has a system of gears which permits the drums to be slowly tilted, as shown in Fig. 4.10, allowing the sand to pour into the pluviator. The sand is pluviated into the chamber in 5 in. thick lifts.

The distribution of the dry density within the pressure chamber was studied using the density pan method (Trautmann *et al.*, 1985). Five density cans of known volume were placed at ten depths within the chamber as shown in Fig. 4.11. After dry sand was pluviated, the cans were excavated and the dry density was determined. The pluviating process resulted in a remarkably uniform density distribution within the pressure chamber in both the vertical and horizontal directions (Fig. 4.12). The average dry density within the chamber is 112 pcf with a 1.37 pcf standard deviation. The uniformity in the density of the deposited sand was greatly aided by the uniform grain size distribution of the pluviated sand.

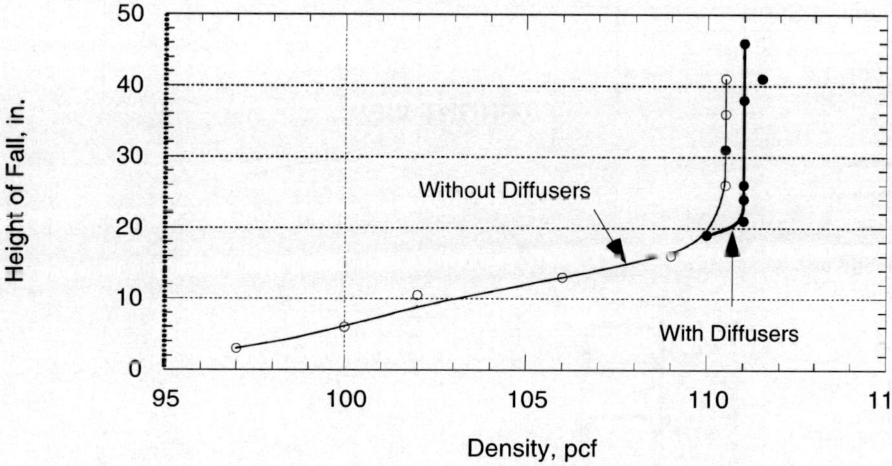

Fig. 4.7 Effect of the Height of Fall on the Dry Density of Sand

[4] East Syracuse, NY 13057.

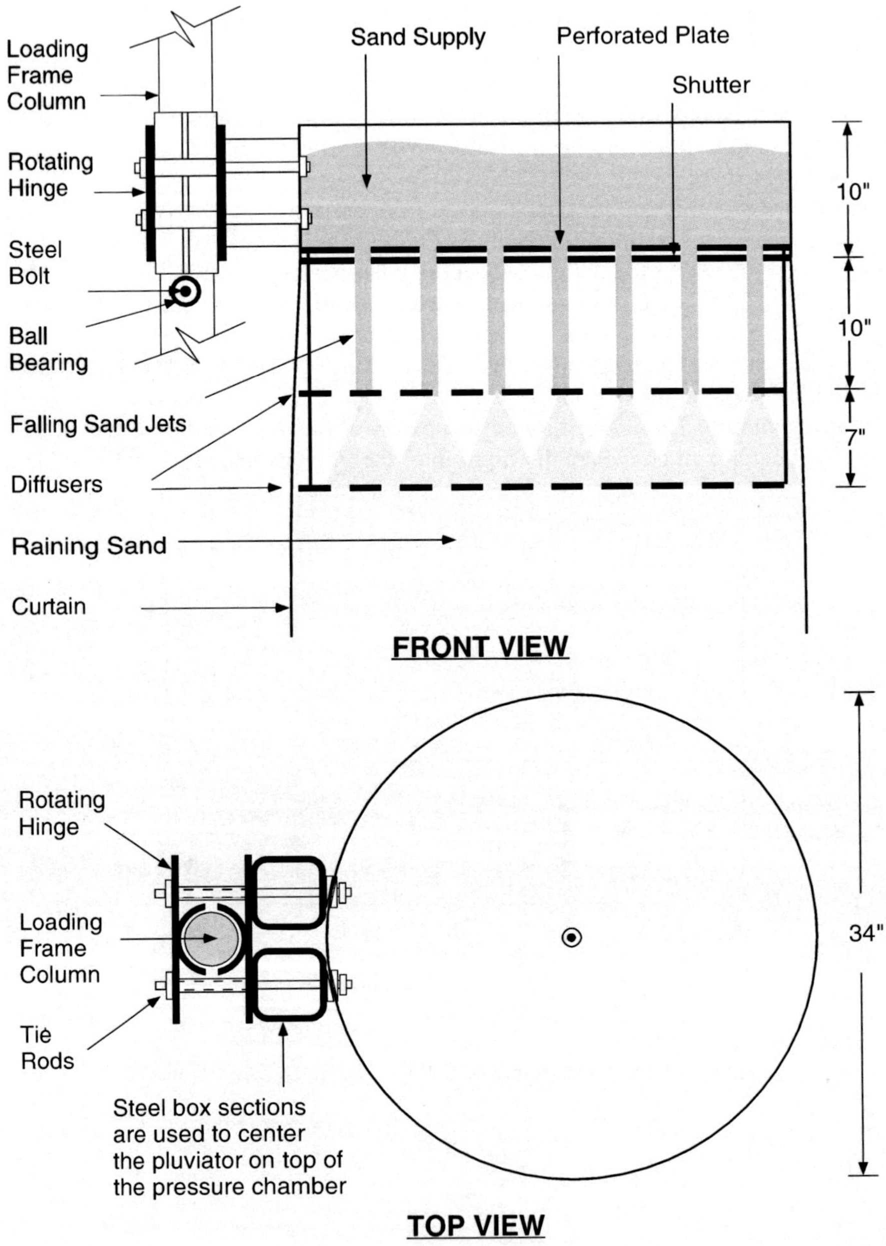

Fig. 4.8 Schematic of the Pluviator

Fig. 4.9 Photograph of Pluviated Sand Passing Through the Diffusers

Fig. 4.10 Photograph of Operator Using the Pluviator

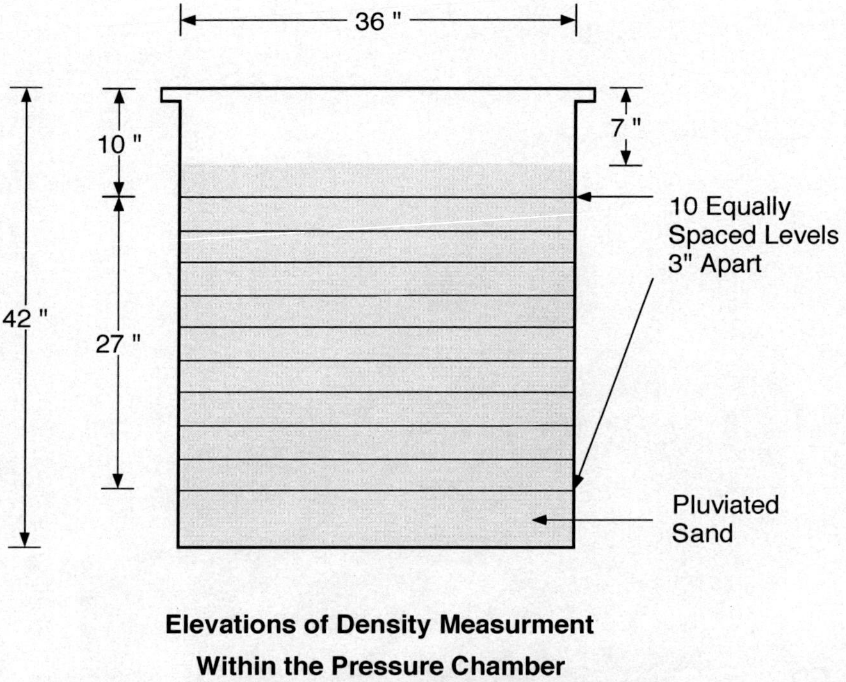

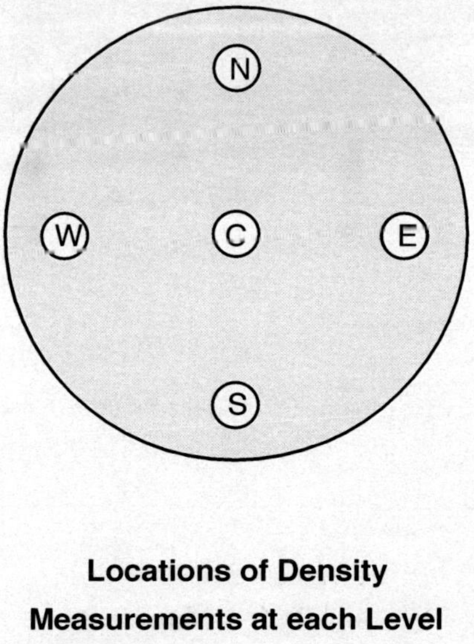

Fig. 4.11 Locations of Density Measurements

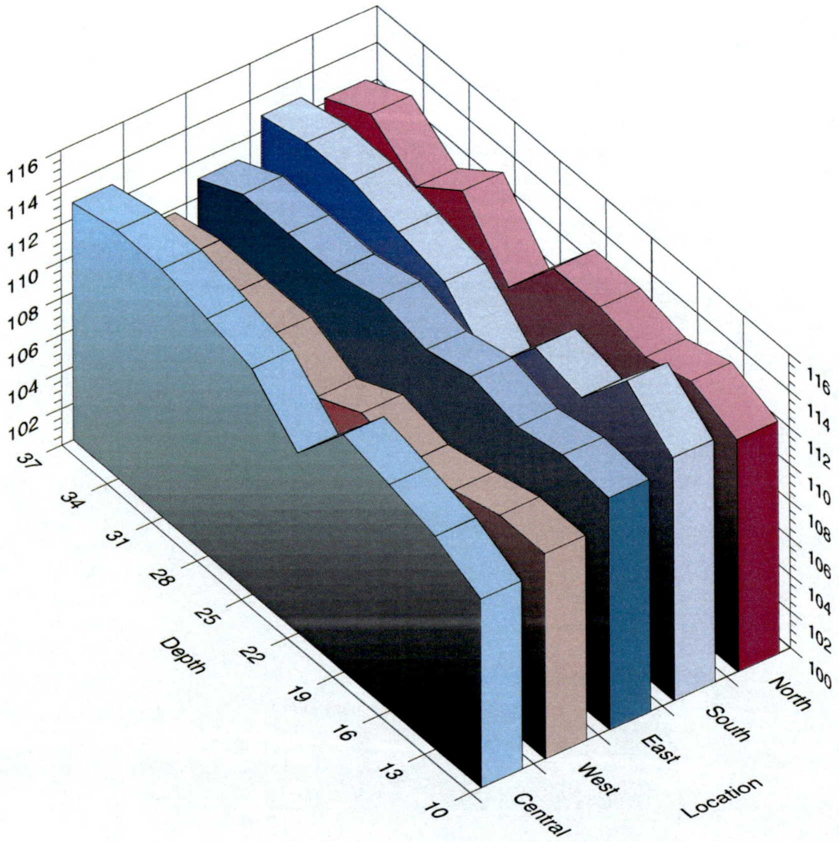

Fig. 4.12 Density Distribution Within the Pressure Chamber

4.3.2.2 Saturation of the Soil Specimen

The water used to saturate the soil specimen is deaired in order to reduce compliance errors in pore pressure measurements. After lateral and vertical pressures are applied, water is introduced at the base of the dry soil sample and allowed to rise slowly. It is important to note that application of the pressure after the sample is saturated results in liquefaction of the soil specimen. A pressure head of 1–1.5 psi is required to saturate the system in 2–3 hours.

A large volume of deaired water is required to saturate the sand. A deairing water system capable of deairing 120 gallons of water in 20 minutes was built. The system shown in Fig. 4.13 and 4.14 consists of a steel pipe, 6 ft. tall, 24 inches in diameter, and 0.25 in. thick. Two plates are welded on the ends of the pipe; the top plate is 0.5 in. thick and the bottom plate is 0.25 in. thick. After fabrication, the system was galvanized to protect it from rust. A 1/3 horsepower[5], 3600 rpm electric motor (Dayton model No: 3K979A) is mounted on the top plate. The electric motor is mounted in a machined housing which seals against the motor using a rubber gasket. The motor's shaft extends through the housing and seals against two automotive oil seals which are press fit into the housing. The housing is bolted to the deairing system's top plate using nine studs which are welded to the top plate and extend upwards. The housing seals against the top plate using a rubber gasket. An impeller is connected to the motor's shaft using an extension rod and a coupling.

The top of the deairing system is connected to a tap water source and a pressure panel board. The same pressure panel board is used to control both the water deairing system and the confining pressure on the pressure chamber. The bottom of the water deairing system is connected through valves to a drain and to the bottom of the pressure chamber. A sight tube indicates the level of water in the water deairing system. After the tank is filled with fresh water, deairing is performed by first connecting the water deairing system to the university shop vacuum which reduces the pressure in the system to approximately 15 in. of mercury. Next, the system is connected to a vacuum pump that further reduces the vacuum to 25 in. of mercury. The electric motor is then turned on and water is spun under a vacuum of 25–27 in. of mercury, which increases the velocity head and decreases the pressure head. As a result, cavitation is induced in the water, and air bubbles are released from solution. As the process continues, air is released and collected by the vacuum pump until the solution becomes deaired. The system can hold a vacuum of 25–27 in. of mercury for a few days and water is stored under vacuum.

The oxygen content of water is proportional to the air content. Measuring the oxygen content of water is however easier than measuring the air content. A calibrated oxygen probe was used to determine the run time required to deair the water in the deairing system.

[5] A larger motor is recommended. This motor can only spin water under vacuum.

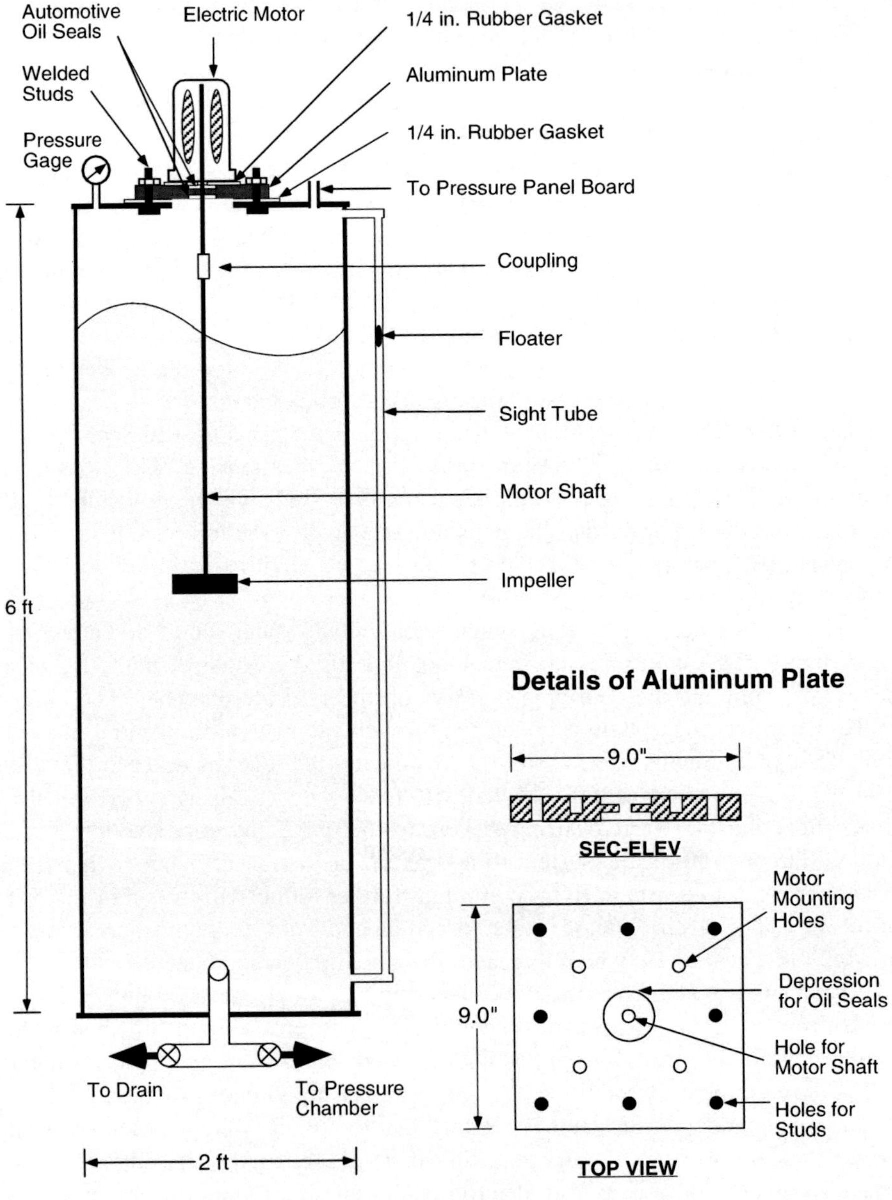

Fig. 4.13 Schematic of the Water Deairing System

Fig. 4.14 Photograph of the Water Deairing System

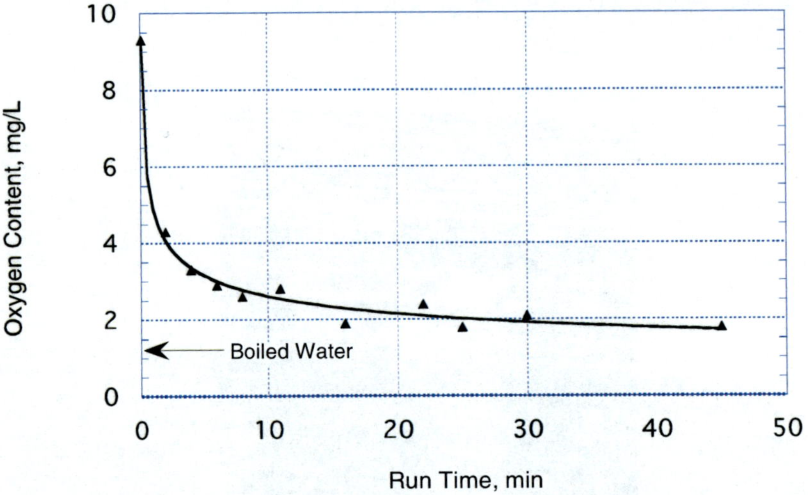

Fig. 4.15 Reduction in Oxygen Content Due to Water Deairing

The deairing system was filled with fresh water and the oxygen content of water samples taken from the system after known intervals of operation was determined. Oxygen content was compared to that of water boiled for 30 minutes. Twenty to thirty minutes of operation are considered sufficient to deair the water (Fig. 4.15).

An important factor in the design of the water deairing system is to ensure that it does not implode when vacuum is applied. Cylindrical vessels subjected to an external pressure, such as the water dearing system, fail in buckling. A stress based design criterion grossly overestimates the capacity of the vessel. Conversely, a stability based design criterion which neglects the stiffness provided by the end plates grossly underestimates the capacity of the vessel. For a cylinder having a Poisson's ratio of 0.3, the critical in-plane buckling load midway between two stiffeners is given by Blake (1990) as:

$$P_c = \frac{0.87 \cdot E \cdot k}{m^{1.5}} \tag{4.1}$$

$$m = \frac{R}{T} \tag{4.2}$$

$$k = \frac{T}{L} \tag{4.3}$$

where:

P_c	external buckling pressure, ksi
E	modulus of elasticity taken as 29,000 ksi
R	mean radius taken as 12 in.
T	wall thickness of cylinder taken as 0.25 in.
L	length of cylinder taken as 72 in.

As designed, the water deairing system can withstand an external pressure of 450 psi, based on Eqs. 4.1-4.3.

4.3.2.3 Sand Drying

At the completion of each test, the soil specimen was excavated and dried in order to prepare it for pluviation into the chamber once more. By far, the most time-consuming aspect of the recycling process was drying the sand. Several methods were tested to determine which was the most efficient. None of the methods was efficient at all water contents and a procedure utilizing drainage, soil vapor extraction, and evaporation was devised as follows:

Drainage: The drainage valves at the bottom of the chamber were opened to drain as much water as possible.

Soil Vapor Extraction: The damp sand was next shoveled into specially prepared 55 gallon drums shown in Figs. 4.16 and 4.17, where initial drying was performed. The inside perimeter of the drums was lined with a geocomposite filter. Each drum was heated with two 1000-watt heat straps[6]. A 3-inch perforated pipe with an electric fan connected to its top was placed at the center of the sand-filled drum. Typically the temperature of the sand was raised to about 150–170˚F, and then the electric fan was turned on. The fan draws air along the walls of the drum and side-ways through the sand. The soil vapor extraction procedure was found to be most efficient during the first two to three days.

Evaporation: Further drying was achieved by emptying the barrels (one at a time) onto the steel floor of the environmental chamber shown in Fig. 4.18. The sand was spread to form a two to three inches thick layer over the 12 ft by 12 ft floor. The environmental room was designed to sustain a constant temperature and humidity, which are set on its control panel. The temperature and humidity were set to 140˚F and zero, respectively. Typically, the water evaporated in 12 to 24 hours, and the dry sand was shoveled into 55 gallon drums to await pluviation into the pressure chamber.

---

[6] Electro Flex Heat, Inc., Bloomfield, Connecticut 06002.

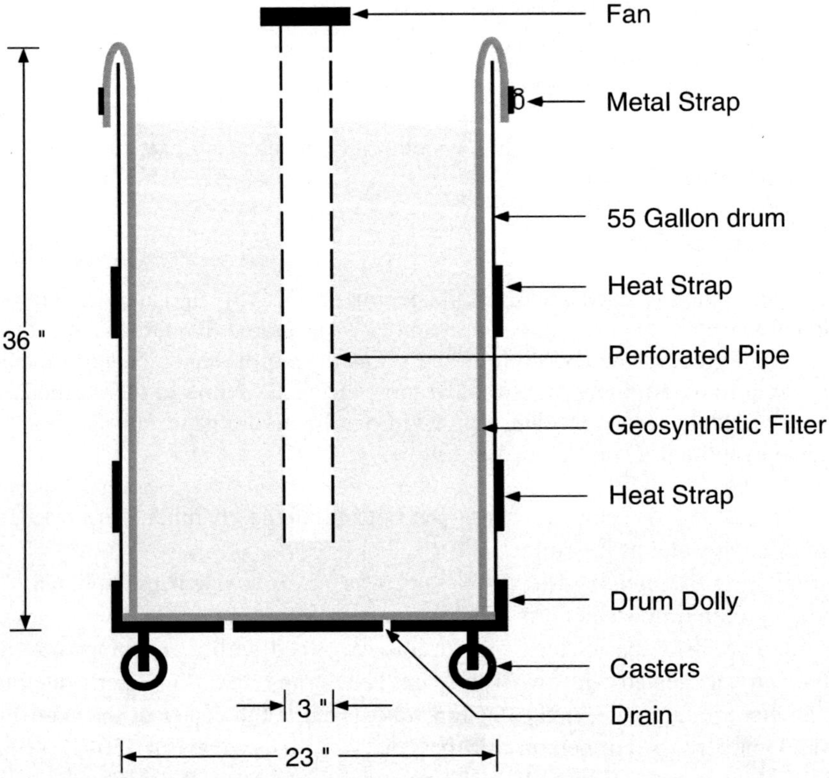

Fig. 4.16 Schematic of the Sand Drying Drum

Fig. 4.17 Photograph of the Sand Drying Drum

Fig. 4.18 Environmental Chamber

4.3.3 Chamber Effects

The finite size of the pressure chamber results in a number of variations from field behavior (Al-Douri *et al.* 1993). These variations may influence the behavior of the piles tested in the chamber and therefore should be carefully considered.

4.3.3.1 Size Effects

The pressure chamber has an internal diameter of 34.8 inches. The piles tested in the pressure chamber are 3.5 in. in diameter. As such, the ratio between the diameter of the chamber to the diameter of the tested piles is approximately 10:1. The diameter ratio is important, because the proximity of the boundary of the chamber may influence the behavior of piles. The pipe piles tested in the chamber have wall thicknesses ranging between 0.125 and 0.31 in. These wall thicknesses correspond to cross sectional areas of closed ended piles having diameter ratios ranging between 20 and 30.

Parkin and Lune (1982), Parkin (1988), and Schnaid and Houlsby (1991), among others, investigated the effects of the diameter ratio on the cone penetration resistance in dense and loose sands. For loose sands ($D_r < 30\%$), the diameter ratio has a negligible effect on the penetration resistance. In dense sands ($D_r > 90\%$), chamber size effects are perceptible at diameter ratios below 50–60. For over-consolidated sands tested under stress controlled boundary, diameter ratio effects are negligible for ratios exceeding 33. It is therefore believed that a small increase in experimental pile capacities may occur due to size effects.

Table 4.1 Boundary Conditions of the Pressure Chamber

	Lateral	Top	Bottom
Stress / Strain	Stress Controlled	Stress Controlled	Strain Controlled
Drainage	Undrained	Drained	Drained

4.3.3.2 Boundary Conditions

The boundary conditions at the edges of the sample may also have an effect on the pile's capacities (Table 4.1). Two different boundary conditions bracket the lower and upper bounds of the free field behavior. These are the controlled stress and controlled strain boundaries (Veismanis 1974). The chamber has a stress controlled boundary which results in a reduction in pile capacities from equivalent field values. Boundary effects and size effects change the measured capacities in opposite directions and may cancel each other out.

4.3.3.3 Aging Effects

The effect of time on the mechanical behavior of clays has long been accepted. In the past 20 years, aging effects on the mechanical properties of sands have also been recognized (Schmertman 1989). Evidence of aging in sands is reported for small strain dynamic tests (Anderson and Stokoe 1978), large strain consolidation tests (Mesri *et al.* 1990), and in field tests (Skempton 1986). Aging effects are attributed to drained creep and cementation (Ghionna and Jamiolkowski 1991).

The piles tested in the pressure chamber are installed in freshly deposited sand, and could therefore exhibit less capacity than equivalent piles in the field due to lack of aging. The evidence is, however, that aging results in an increase in the stiffness of the sand, and is therefore more important for deformational problems. For limit equilibrium type problems, like pile capacity, aging effects are considered secondary to size and boundary effects.

4.3.3.4 Wave Reflection Effects

Ideally, the boundaries of the pressure chamber should absorb pile driving stress wave energy with no reflections. Wong (1985) estimated that radiation damping in The University of Houston LVLPSC chamber to be on the order of 10% of the critical value. The estimated value is thought to be equivalent to that obtained for a pile embedded in a half elastic space (O'Neill, 1991). The LVLPSC chamber, which is 30 in. in diameter and 100 in. long, is smaller in diameter than the pressure chamber described in this chapter. Accordingly, the energy absorbing characteristics of this pressure chamber are considered acceptable.

4.4 The Loading System

Static and axial loads are applied to the piles using a loading system which consists of a loading frame and a feedback control system. The loading system is designed to apply dynamic loads, without transferring any vibrations to the building structure.

4.4.1 The Loading Frame

The loading frame has a dynamic design capacity of 60 kips. The frame consists of two 4-in. diameter steel columns, 48 in. apart, which are bolted to two 12-in. beams to form a box frame (Fig. 4.19). The tips of the columns are placed inside four shoes which are fixed to pre-existing rails in the ceiling and floor of the laboratory. The box design prevents the transfer of any loads to the structural frame of the building; thus preventing direct transfer of vibrations to the top floors.

The height of the loading frame was determined by the available ceiling height which is 14 ft. The elevation of the cross head beam (top beam) is adjustable to facilitate research on other types of foundations such as footings. The cross head is currently being used at its maximum elevation which is 139 in. above the floor level. The lower beam is fixed in its position 26 in. above the floor. The available clearance is 113 in. high, by 44 in. wide.

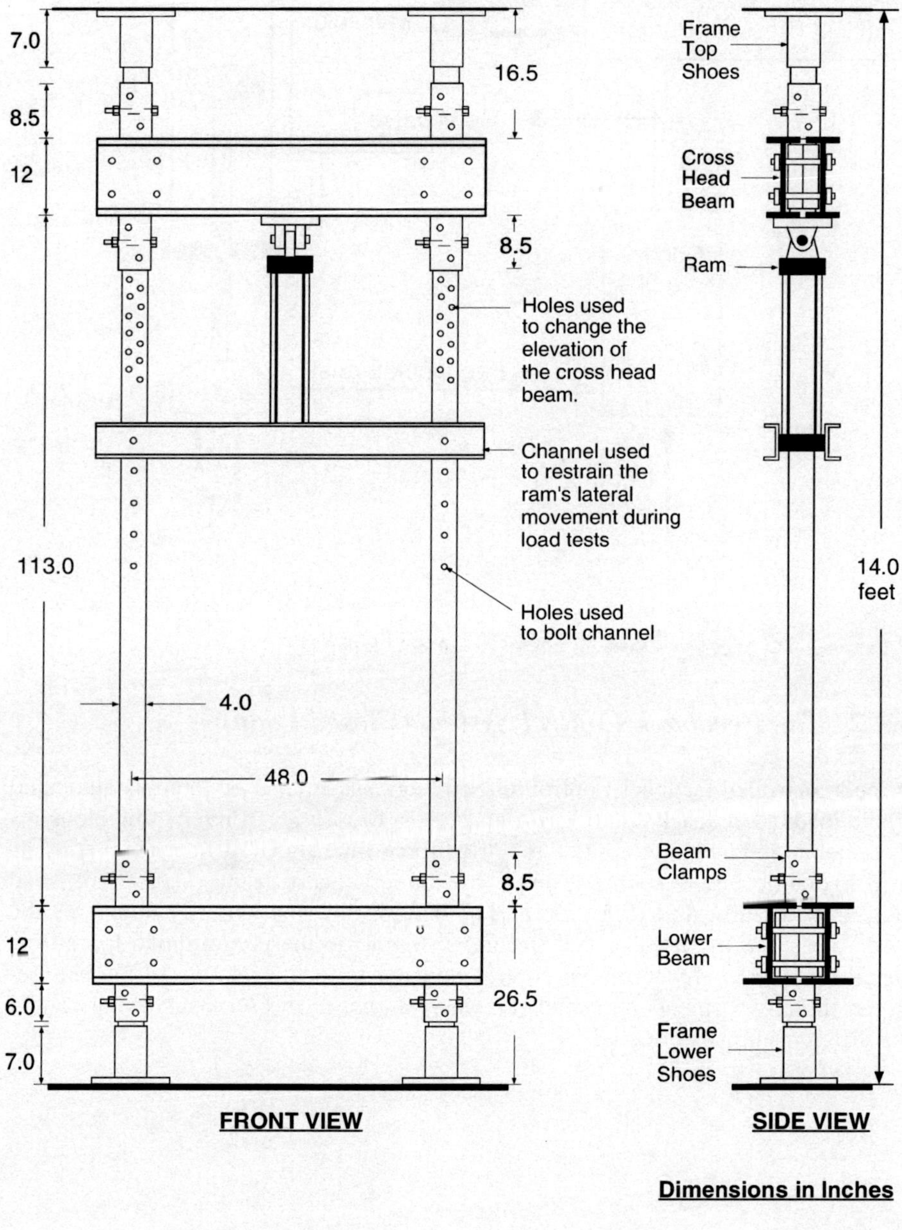

Fig. 4.19 Details of the Loading Frame

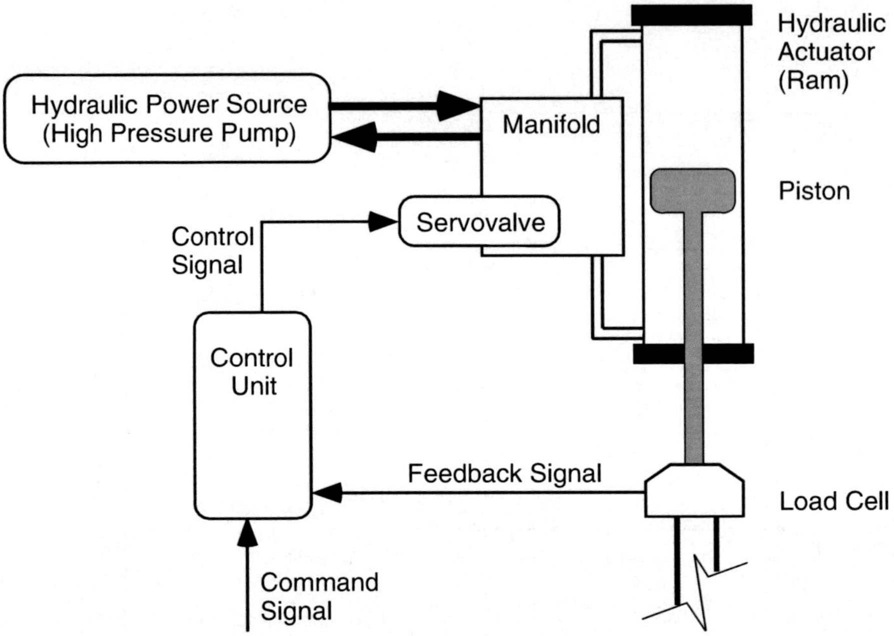

Fig. 4.20 The MTS Closed Loop System

4.4.2 The Feedback Control System (Closed Loop)

A load-controlled feedback control (closed loop) system is used to apply static and cyclic loads. Feedback control systems consist of a series of interacting elements that operate in tandem to apply a specified force-time signal. The system consists of a hydraulic power source (pump), hydraulic actuator (ram), servovalve, load cell, and a control unit (Fig. 4.20). The control unit continuously compares the loads measured by the load cell (*feedback signal*) to the programmed force-time signal (*command signal*). The control unit reacts to the relative difference between these two signals and sends a *control signal* to the servovalve that causes the difference to be corrected.

4.4.2.1 The Hydraulic Power Source (Hydraulic Pump)

The hydraulic power source is the source of hydraulic fluid under pressure. The pump (MTS model No: 510.21B) can provide pressures up to 5000 psi. The pump is regulated to provide a pressure of 3000 psi during load tests and a pressure of 500 psi during test set up. The pump has a flow capacity of 21 gpm.

4.4.2.2 The Hydraulic Actuator (Hydraulic Ram)

The hydraulic actuator (hydraulic ram) is the force generating device in the system. The hydraulic actuator consists of a cylinder with a piston connected to a piston rod. Fluid pressure is applied to one side of the piston, causing it to move. When the piston rod encounters an external reaction, a force equal to the effective piston area multiplied by the actuating pressure is generated. At the 3000 psi operating pressure, the hydraulic actuator (Miller model No: H-84 W30190H) has a maximum compressive capacity of 58.9 kips, and a maximum tensile capacity of 44.2 kips. The maximum stroke is 24 in. and the diameter of the piston rod is 2.5 inches.

The hydraulic actuator is connected to the cross head of the loading frame using a swivel connection which allows the ram to be rotated away from the loading frame during sand pluviation or pile driving (Fig 4.2). During loading, the actuator is fixed in a vertical position using two aluminum channels in order to prevent eccentric loading (Fig 4.19). The actuator hydraulic connection was kept short, such that it must be disconnected when the ram was in the rotated position, in order to prevent accidental powering of the ram in the rotated position.

4.4.2.3 The Servovalve

The servovalve controls the hydraulic actuator by opening or closing in response to a control signal from the control unit. The servovalve can open in either of two directions allowing high-pressure fluid to flow into either side of the hydraulic actuator. When the servovalve is opened to allow fluid flow into one end of the cylinder, it provides a path for fluid to flow from the other end of the cylinder back to the hydraulic pump. The servovalve is a proportional device which allows a rate of fluid flow in direct proportion to the magnitude of the control signal. The polarity of the control signal determines which end of the actuator cylinder receives additional fluid, and therefore determines the direction of the piston movement and applied force.

4.4.2.4 The Load Cell

A 50,000 lb load cell (Interface model No: 1220 AMS) is the feedback transducer of the MTS system. The load cell consists of two electrically isolated bridges with each bridge consisting of 24 strain gages. One bridge senses the feedback signal, while the other bridge is used to record the loads on the data acquisition system. The two bridges should measure the same load. It was not possible to use a single-bridge load cell to accomplish both functions due to ground loops.

The nominal excitation voltage of both load cell bridges is 10 V. Each bridge is, however, powered and conditioned in a different fashion. The MTS control unit provides excitation and signal conditioning for the bridge which senses the feedback signal. A separate power source and signal conditioning unit are used to power and condition the bridge connected to the data acquisition system.

4.4.2.5 The Control Unit

The control unit consists of the following components:

- master control panel (MTS model No: 413.05)
- controller (MTS model No: 442.11) with the following attachments:
  - servo controller (MTS model No: 440.13)
  - valve driver (MTS model No: 440.14)
  - two DC signal conditioning units (MTS model No: 440.21)
  - AC signal conditioning unit (MTS model No: 440.22)
  - amplitude measuring unit (MTS model No: 440.51).
  - limit detector (MTS model No: 440.41)
  - feedback selector (MTS model No: 440.31)
- counter panel (MTS model No: 717.01)
- function generator (MTS model No: 410.31)

The operating range of the MTS system is determined by two factors, the load cell capacity (50,000 lb), and the setting of a range switch, which may be set to 100%, 50%, 20%, or 10% of the load cell capacity. The range switch is set to 50%, and the MTS system is set to operate in a ± 25,000 lb range. The reference voltage of the controller is ± 10 V, which means that the controller uses a scale of ± 10 V to represent loads in the range of ± 25,000 lb. The resolution of the control load depends on the setting of the range switch and is 50 lb for the selected setting. The command signal is programmed as the sum of a constant voltage referred to as the *set point*, and the product of the function generator output ranging between ± 10 V, multiplied by a *span* constant. The combination of the function generator, set point, and span can be used to represent a wide range of force-time loading paths. The control unit compares the feedback signal to the command signal and provides an *error signal* which is proportional to the difference between the two inputs. The error signal is further conditioned to generate the control signal which is fed to the servovalve. Optimal performance is achieved using a multitude of additional adjustments including, limit detectors, gain control, dither control, counters, error detectors, HPS (pump) control and pressure regulation, and various voltage adjustments.

4.5 The Data Acquisition System

The data acquisition system is capable of sampling up to 64 channels and has a maximum sampling frequency of 1 MHz. Signal conditioning is provided for 32 channels which are connected to a load cell, LVDT, position transducer, strain gages, and pore pressure transducers (Table 4.2). The system consists of a computer, an analog to digital converter, Laboratory Workbench data acquisition software, sensors, and signal conditioning equipment (Fig. 4.21).

4.5.1 The Computer

The heart of the data acquisition system is Concurrent Computer Coorporation MASSCOMP 5400 workstation. Concurrent computers are designed for high performance real-time applications. The computer achieves balanced performance without bottlenecks by off-loading many functions from the central processor unit (CPU) to coprocessors which carry out graphics, data acquisition, floating point, and vector arithmetic. The computer uses a modified version of the standard unix operating system. The modification is necessary because unix is a time sharing operating system, while data acquisition is a real time task (Cole and Sundman, 1985). The modification prevents interruption of real time jobs and locks critical data acquisition routines in primary memory. The computer is connected to the local area network (LAN) Ethernet, thus permitting remote login.

4.5.2 Data Acquisition Hardware

Data acquisition tasks are controlled by a dedicated data acquisition processor, which ensures high throughput with few bottlenecks. The attached hardware consists of:

- *AD-12 F Analog to Digital Converter.* The board can sample up to 16 single ended (8 differential) channels with a resolution of 12 bits at rates up to 1 MHz. The system is configured to operate in a bipolar range of ± 5 V.
- *MX-48F Multiplexer.* The board expands the input channel capacity of the AD-12F to 64 single-ended or 32 differential channels.
- *CK-10 Programmable Clock Interface Board.* The board has eight independent clocks, which are programmed to synchronize data acquisition functions.

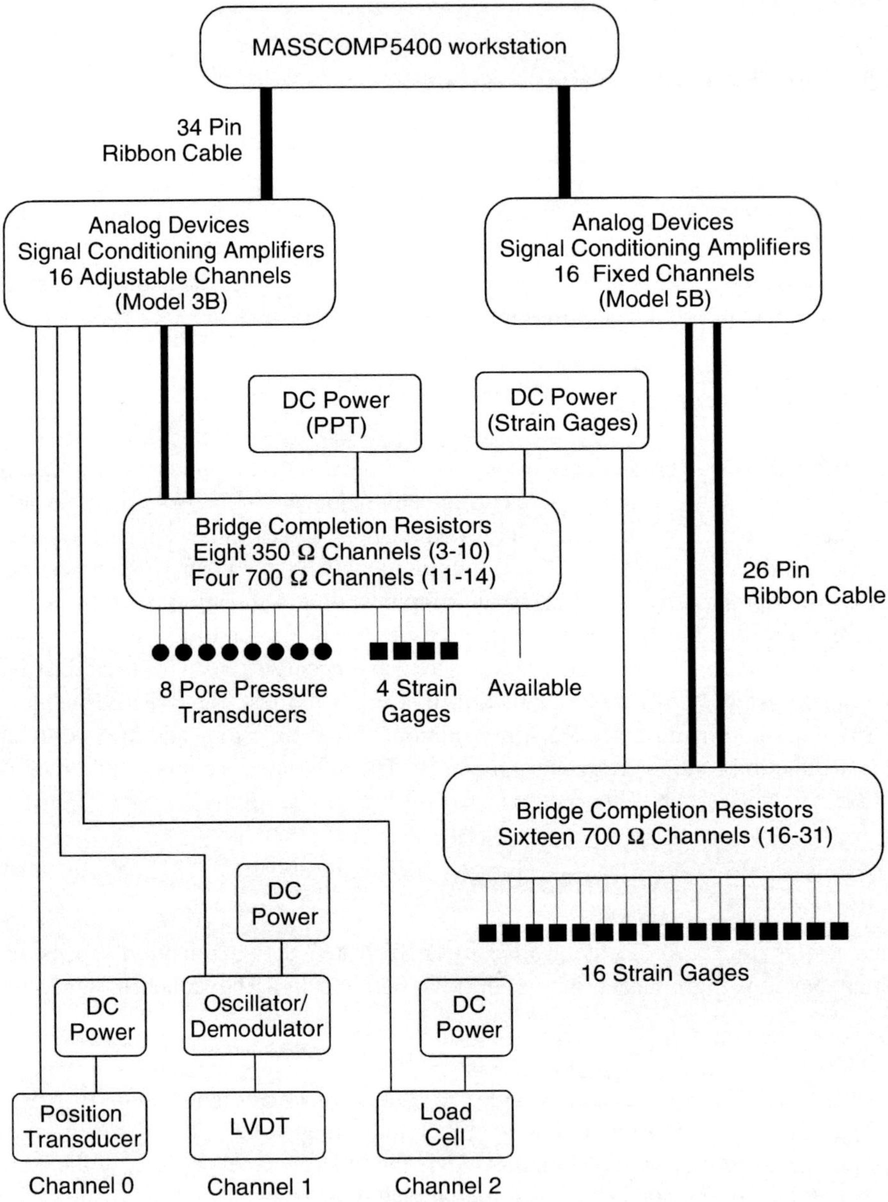

Fig. 4.21 The Data Acquisition System

Table 4.2 Specifications of The Data Acquisition System

Channel		Location		Gain		Excitation
No:	Description	In/Out	Ht.	Nominal	Type	(Volt)
0	Position Transducer	...	...	1	Adj.†	20
1	Load Cell	...	...	200	Adj.	10
2	LVDT	...	...	1	Adj.	AC
3	Pore Pressure Transducer	Out	2	1000	Adj.	5
4	Pore Pressure Transducer	Out	5	1000	Adj.	5
5	Pore Pressure Transducer	Out	17	1000	Adj.	5
6	Pore Pressure Transducer	Out	10	1000	Adj.	5
7	Pore Pressure Transducer	In	2	1000	Adj.	5
8	Pore Pressure Transducer	In	17	1000	Adj.	5
9	Pore Pressure Transducer	In	5	1000	Adj.	5
10	Pore Pressure Transducer	In	10	1000	Adj.	5
11	Axial Strain Gage	In	2	1000	Adj.	10
12	Circumfrential Strain Gage	In	2	1000	Adj.	10
13	Axial Strain Gage	In	5	1000	Adj.	10
14	Circumfrential Strain Gage	In	5	1000	Adj.	10
15	Available (Programmable)	...	...	prog.	Adj.	...
16	Axial Strain Gage	In	10	500	Fixed	10
17	Circumfrential Strain Gage	In	10	500	Fixed	10
18	Axial Strain Gage	In	17	500	Fixed	10
19	Circumfrential Strain Gage	In	17	500	Fixed	10
20	Axial Strain Gage	In	24	500	Fixed	10
21	Circumfrential Strain Gage	In	24	500	Fixed	10
22	Axial Strain Gage	Out	2	500	Fixed	10
23	Circumfrential Strain Gage	Out	2	500	Fixed	10
24	Axial Strain Gage	Out	5	500	Fixed	10
25	Circumfrential Strain Gage	Out	5	500	Fixed	10
26	Axial Strain Gage	Out	10	500	Fixed	10
27	Circumfrential Strain Gage	Out	10	500	Fixed	10
28	Axial Strain Gage	Out	17	500	Fixed	10
29	Circumfrential Strain Gage	Out	17	500	Fixed	10
30	Axial Strain Gage	Out	24	500	Fixed	10
31	Circumfrential Strain Gage	Out	24	500	Fixed	10

‡ Location refers to the location on the double-wall pile. Height refers to the height above the tip of the double wall pile (Fig. 5.11).

† Adjustable gain and span (5–10% range).

4.5.3 Data Acquisition Software

Data acquisition programming is performed graphically using Laboratory Workbench software. The software consists of a number of modules which represent common data acquisition tasks, such as analog input, storage, multiplexing, de-multiplexing, on-off switching, triggering... etc. A *virtual instrument* is programmed by connecting the desired modules graphically on the screen. The software also supports limited signal processing routines such as fast fourier transforms (FFT/IFFT), filters, and power spectrums. The data may be displayed in real-time on virtual digital voltmeters, oscilloscopes, or X-Y plotters. The virtual instrument used to display and record load test results is shown in Fig. 4.22.

Each module has a large number of settings which must be entered. These settings are stored in separate files. A handy feature of the software is that virtual instruments may be used with more than one setting file. For example, the same instrument is used to record the results of load tests on instrumented piles with 31 sensors and un-instrumented piles with three sensors. The setting files for the two tests specify the number of channels to be recorded as well as the various settings for each channel and module used.

Laboratory Workbench, together with the 8 clocks of the CK-10 clock interface board provide tremendous control over timing sequences. Data may be sampled continuously, or periodically in sweeps, or in frames within sweeps. Periodic sampling may be triggered externally or internally, in response to an external event, or at regular intervals. Periodic sampling is a powerful tool for capturing dynamic phenomena such as pile driving. A large sampling frequency is required to describe a dynamic wave form such as the force-time record of the pile head subsequent to a pile hammer blow. The combination of periodic sampling and internal triggering facilitates sampling at a high frequency at the frames of interest only, thereby relieving any bottlenecks.

4.5.4 Signal Conditioning Equipment

Signal conditioning equipment was used to amplify, filter, and isolate the output signals of all sensors. Amplification of read-out signals is required in order to increase the resolution of instruments and/or increase the voltage of the readout signal to a level that could be read by the analog-to-digital converter. The LVDT was provided with its own oscillator/demodulator unit, which provided the necessary signal conditioning. All other sensors were conditioned as discussed next.

Filtering is used to eliminate high frequency noise from the recorded signal. A 10 kHz hardware filter was used on all the channels except the LVDT. The sensors are isolated from the computer in order to eliminate ground loops, and improve the safety and reliability of the system.

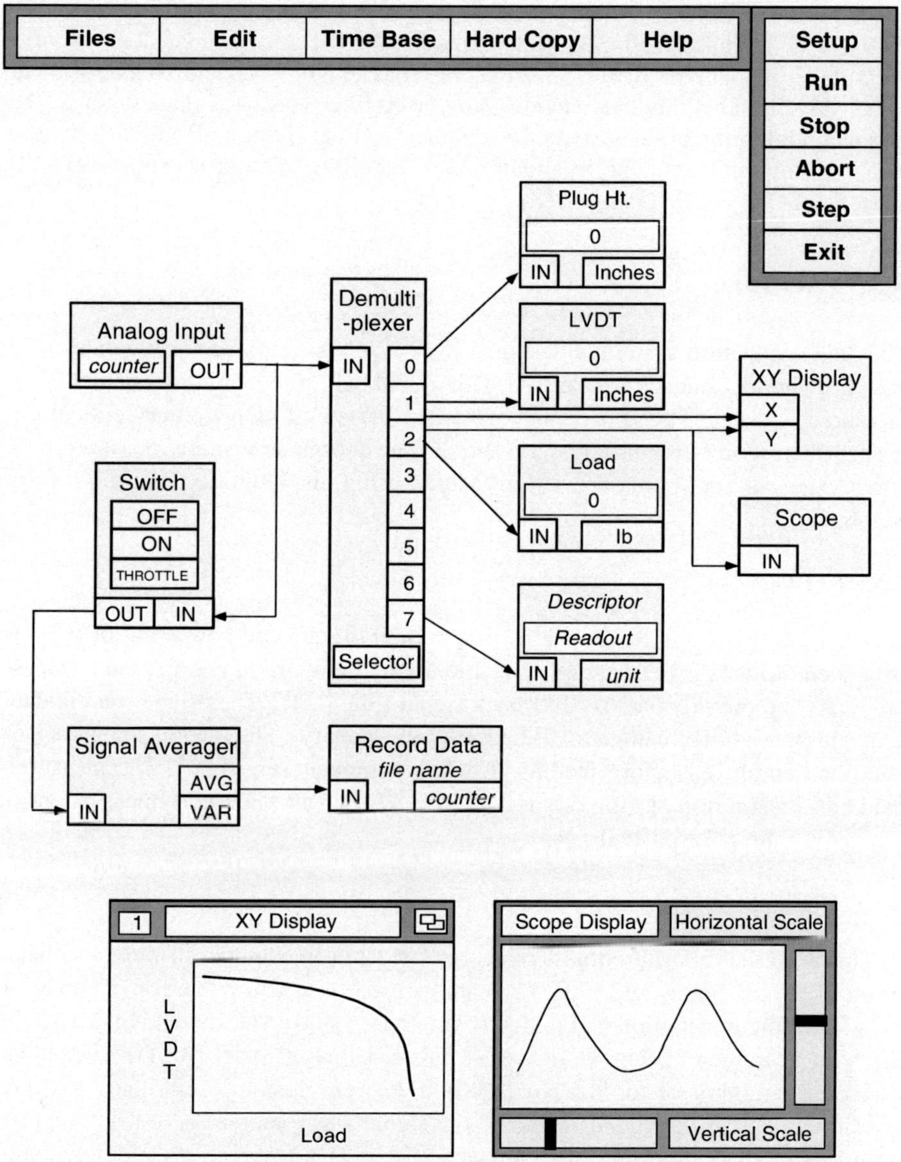

Fig. 4.22 Schematic of Virtual Load Test Instrument

Two signal conditioning units were assembled from components manufactured by Analog Devices. The first unit provided adjustable zero and span controls, and was used for pore pressure transducers, load cells, LVDT, position transducers and strain gages. The second unit had fixed zero and span, and was used for conditioning strain gage signals only. The first unit consists of a back plane (model No: 3B01), power supply (model No: AC1307), and 15 independent signal conditioning modules (model No: 3B40-01 (12 modules), 3B40-02 (1 module), 3B40-00 (1 module), and 3B41-02 (1 module)). The second unit consists of a back-plane (model No: 5B01), power supply (model No: 977), and 16 independent signal conditioning modules (model No: 5B40-01). The back-planes house the signal conditioning modules and provides them with excitation. Both units feature 0.05% calibration accuracy, less than 0.02% non linearity and less than ±1μv/°C thermal drift.

4.5.5 *Sensors*

The data acquisition system is designed to read loads, displacements, strains, and pore pressures. Loads are measured with a load cell, pile head displacements are measured with an LVDT, plug movements are measured with a rotary optical encoder (referred to herein after as a position transducer), strains are measured with strain gages and pore pressures are measured using diaphragm-type pore pressure transducers.

4.5.5.1 Load

As discussed earlier in Section 4.4.2.4, a 50,000 lb load cell consisting of two independent bridges is permanently fixed to the hydraulic ram, and is used to measure loads and provide the MTS feedback signal (Fig 4.23). The bridge used in data acquisition is excited using an HP 6205B power supply. The readout signal is isolated and amplified before feeding it to the computer (Fig 4.21). The measured load has a resolution of 12.2 lb and less than 0.05%-full scale non-linearity (maximum absolute error of 25 lb).

4.5.5.2 Displacement

A Linear Variable Differential Transformer (LVDT) is clamped to the cross head beam of the loading frame (Fig. 4.23) and is used to measure the displacement of the pile during installation and pullout. The LVDT (TRANSTEK model No: 223-000) has a linear range of 30 in. A TRANSTEK power supply (model No: D15.2) and an oscillator/demodulator (model No: 1000-0011) were used to excite the LVDT and condition the read-out signal. The read-out signal was routed back to the computer through signal conditioning box number 1 (Fig 4.21); however, no additional signal conditioning is performed. The LVDT signal has a resolution of 0.0073 in., and less than 0.025%-full scale non-linearity (maximum absolute error of 0.075 in.).

4.5.5.3 Plug Movement

A position transducer (MagneTek model No: P-50-B (A68)) is used to measure the elevation of the soil inside the pile during installation. The transducer consists of a cable wound around a spring loaded shaft which is connected to a rotary optical encoder. When the cable is pulled out of the transducer, the shaft rotates and the magnitude of the rotation is registered by the optical encoder. The transducer has a simulated wheatstone bridge circuit which converts the cable displacement to voltage changes with a linearity better than 0.1% full scale (0.05 in.). The position transducer has a linear range of 50 in. and a resolution of 0.003 inches.

The position transducer was customized to meet the following specifications:

- Ruggedized to withstand 2000G,
- Heavy duty dual parallel cables,
- Water tight,
- Cable acceleration up to 100g.

A 5.6-lb force is required to overcome the resistance of the spring loaded shaft and extend the cable out of the transducer. The free end of the cable is connected to a 5.85-lb brass weight which rests on top of the soil core (Fig 4.24). The net force acting on the soil core is 0.25 lb. Any motion of the core leads to an extension or retraction of the cable and the rotation of the spring-loaded shaft, which is registered by the circuit as a voltage change.

The transducer is excited by a 20 V DC power source (HP model No: 6205B), and the read-out signal is routed to the computer through signal conditioning box number 1, where it is isolated.

4.5.5.4 Strains & Pore Pressures

Strains and pore pressures are measured using the double-wall pile (Chapter 5). The pile is instrumented with 40 strain gages and eight pore pressure transducers which are capable of defining the effective state of stress around it. Two digital power sources (Lambda model No: LJ 4018) are used to provide 10 V and 5 V of excitation for the strain gages and pore pressure transducers, respectively. The read-out signals are first completed using custom-made bridge completion boxes (Section 5.4.4). Next, the signals are amplified and isolated before routing them to the computer (Fig. 4.21). The strain gages and pore pressure transducers have resolutions of approximately 22 lb and 0.01 psi, respectively.

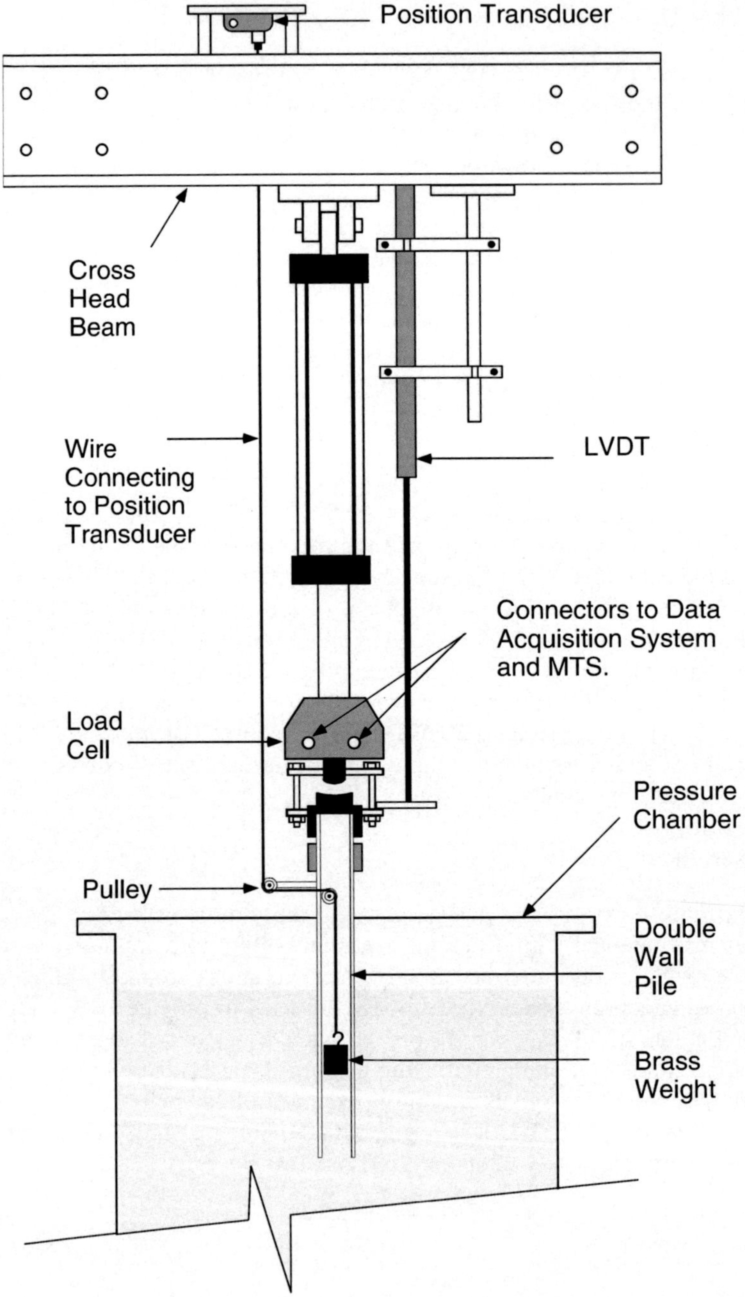

Fig. 4.23 Instrumentation for Load, Displacement, and Plug Movement

4.6 Conclusions

A detailed blueprint for designing an experimental facility consisting of a pressure (calibration) chamber, loading system, and data acquisition apparatus has been provided.

References

Al-Douri, R.H., Hull, T.S., Poulos, H.G.: Influence of Test Chamber Boundary Conditions on Sand Bed Response. Geotechnical Testing Journal, GTJODJ 16(4), 550–562 (1993)

Anderson, D.G., Stokoe, K.H.: Shear Modulus a Time Dependent Soil Property. Dynamic Geotechnical Testing, ASTM, 66–90 (1978)

Blake, A.: Practical Stress Analysis in Engineering Design, 2nd edn. Marcel Dekker, Inc., NY (1990)

Cole, C.T., Sundman, J.: UNIX in Real Time, What it Takes to Make the Grade, UNIX Review (1985)

Ghionna, V.N., Jamiolkowski, M.: A Critical Appraisal of Calibration Testing in Sands. In: Proc. of the 1st Int. Symposium on Calibration Chamber Testing, Potsdam, NY, pp. 13–39 (1991)

Fouse, J.L.: Group Behavior of Axially-Loaded Piles in Sand, Master Thesis, The University of Texas at Austin (1984)

Mesri, G., Feng, T.W., Benak, J.M.: Post Densification Penetration Resistance of Clean Sands. Journal of Geotechnical Engineering, ASCE (1990)

O'Neill, M.: Houston's Calibration Chamber, Case Histories. In: Proc. 1st Int. Symposium on Calibration Chamber Testing, Potsdam, NY, pp. 13–39 (1991)

Parkin, A.K., Lunne, T.: Boundary Effects in the Laboratory Calibration of a Cone Penetrometer in Sand. In: Proc. 2nd European Symposium on Penetration Testing, vol. 2, pp. 761–768 (1982)

Parkin, A.K.: The Calibration of Cone Penetrometers. In: deRuiter, J. (ed.) Proc. Penetration Testing 1988, Balkema, Rotterdam, vol. 1, pp. 221–243 (1988)

Rad, N.S., Tumay, M.T.: Factors Affecting Sand Specimen Preparation by Raining. Geotechnical Testing Journal, ASTM 10(1), 31–37 (1987)

Schmertmann, J.H.: The Mechanical Aging of Soils, 25th Terzaghi Lecture, ASCE (1989)

Schnaid, F., Houlsby, G.T.: An Assessment of Chamber Size Effects in the Calibration of In Situ Tests in Sand. Geotechnique 41(3), 437–445 (1991)

Schofield, A.N.: Cambridge Geotechnical Centrifuge Operations, 20th Rankine Lecture. Geotechnique 30(3), 227–268 (1980)

Skempton, A.W.: Standard Penetration Test Procedures and the Effects in Sand of Overburn Pressure, Relative Density, Particle Size, Aging, Over consolidation, Geotechnique, No. 3 (1986)

Trautmann, C.H., Kulhawy, F.H., O'Rourke, T.D.: Sand Density Measurements for Laboratory Studies. Geotechnical Testing Journal 8(4), 159–165 (1985)

Veismanis, A.: Laboratory Investigations of Electrical Friction Cone Penetrometers in Sand. In: Proc. European Symposium on Penetration Testing, Stockholm, vol. 2, pp. 407–420 (1974)

Wong, D.O.: Design and Analysis of an Apparatus to Simulate Density Stresses in Deep Deposits of Granular Soils, Ph.D. Thesis, University of Houston (1985)

6. Electro Pneumatic Laboratory Pile Hammer

Abstract. This chapter presents the design, manufacturing, and performance of a single-acting electro-pneumatic pile hammer, suitable for installing model piles in a laboratory calibration chamber. The performance of the hammer was evaluated by instrumenting the various components of the hammer-pile system and monitoring critical parameters such as pressure, displacement, force, and velocity. The hammer delivers a rated energy of 156 ft-lb with a 66% efficiency. The hammer can operate with the full design energy at frequencies up to 1.2 Hz, and has been used successfully to model the pile driving process in the laboratory. Sufficient details are provided so that readers can build similar hammers to suit their needs.

6.1 Introduction

One of the major obstacles to an improved understanding of the behavior of piles has been the scarcity of full-scale load test data on instrumented piles. The costs associated with full-scale load tests sway most pile foundations research from field studies to numerical and laboratory model studies. A necessary ingredient in modeling the behavior of piles is modeling the field driving process. Early laboratory studies on piles did not model the driving process, and piles were usually buried or pushed into place, *e.g.* Vesic (1967). The results of more recent studies indicate that the dynamic interaction between the pile and the surrounding soils may greatly influence the state of stress and soil conditions in the vicinity of the pile (Nunez *et al.* 1988). These changes are likely to govern the physical processes that control the behavior of piles during static and cyclic loading.

Despite the importance of modeling the pile driving process, there is little discussion in the mainstream literature regarding the design of laboratory pile hammers. This chapter addresses this need by providing a blue print for designing and building laboratory pile hammers.

6.2 Background

Pile foundations have been used for thousands of years. Throughout these years driving has been the preferred method of installing piles. The Greek historian and traveler Herodotus, who lived in the fourth century B.C., recorded that African tribes lived in dwellings erected on piles that were manually driven into the

M. Iskander: Behavior of Pipe Piles in Sand, SSGG, pp. 127–176.
springerlink.com

lakebed (Fleming *et al.* 1985). Mechanical pile driving came into existence in 1802 when the first steam-powered hammer was used to drive piles in the London docks (Fleming *et al.* 1985). Today, with the advancements in civil engineering, it is required to install piles that are longer, larger in diameter, and have much higher capacity than piles a few decades ago. Pile hammers have evolved and their capacities have kept up with the required pile capacities. Today, production pile hammers have energies reaching 22,00 kip ft and operate at frequencies ranging between 0.6–2 Hz.

Pile design methods have historically been experience-based. Even today, design standards such as API RP-2A are based on experience (Pelletier *et al.* 1993). However, understanding the physical mechanisms that control pile behavior is becoming more important in order to design piles with much higher capacities than the experience base of the civil engineering community, particularly in offshore environments. Ideally, such understanding should be obtained from full-scale load tests on instrumented piles. However, the costs associated with running such tests may exceed a few million dollars per test, particularly for large offshore piles. As a result, most pile research is done with numerical or laboratory model studies.

A necessary ingredient in modeling the behavior of piles in the laboratory is modeling the pile driving process. The University of Houston (UOH) pile hammer was the first automatic laboratory model pile hammer. The UOH hammer consists of a 470 lb ram which free falls from a height of 30 in. at a rate of 0.33 Hz (Ugaz 1988). The hammer is single-acting, air operated, and has an efficiency of approximately 50%. There is also a number of manual laboratory drop hammers that are used in foundations research such as the hammer used by Bernardes (1989) at the Norwegian Institute of Technology (NTH). The use of a slow or manual pile hammer limits the ability to model the driving process. Two types of mechanisms occur when piles are driven, dynamic and diffusive mechanisms. Dynamic effects deal with the inertial forces developed in the soil to resist pile penetration. Diffusive effects deal with the build-up and dissipation of pore water pressures during driving. Slow hammers are capable of modeling the dynamic effects that occur during pile driving but not the diffusive effects.

A number of model pile hammers have also been built for use in a centrifuges For example hammers were built at The University of Colorado (Mehle 1989), The University of Florida (Gills 1988), California Institute of Technology (Allard 1990), The University of Manchester (Craig 1985), and Cambridge University (Nunez *et al.* 1988). Centrifuge hammers are fast but have tiny rams and strokes. For example, The University of Colorado pile hammer operates at a 25 Hz frequency, but has a ram that weighs 0.37 lb and has a stroke of 0.47 inches. Pile driving in a centrifuge is greatly aided, however, by the increased gravitational field of the centrifuge.

One of the objectives of this study is to build a pile hammer that meets the following design criteria:

- The hammer must be sufficiently fast to correctly model both the dynamic and diffusive effects, which are associated with pile driving.
- The hammer must be capable of delivering sufficient energy to install piles in pressurized calibration chamber.
- The length of the hammer and its guide system must not exceed 40 in., which is the available head clearance in the lab.
- Noise, pollution, and vibration must be limited to a level consistent with operating indoors.
- The hammer must be sufficiently robust to withstand many hours of pile driving.
- The hammer must be sufficiently light to facilitate manual setup.

6.3 Special Considerations in the Design of Laboratory Pile Hammers

The design of laboratory pile hammers for use with pressure or calibration chambers poses a number of challenges that are peculiar to the laboratory environment. The effects of the laboratory operating conditions on the design, geometry, and type of laboratory pile hammers are discussed next.

6.3.1 Type of Hammer

Pile hammers may be classified into two main types, internal combustion and external combustion. Internal combustion hammers are diesel operated, while external combustion hammers are powered pneumatically or hydraulically. The two most common methods of operation are termed single-acting and double-acting. A single-acting hammer is operated by lifting the ram to the desired height before releasing it to fall freely. A double-acting hammer operates in a similar fashion except that the ram is powered in both the down and up-strokes. There are advantages and disadvantages associated with the use of different types of hammers.

The selection of the laboratory hammer's operating principle involves a process of elimination. Diesel hammers were eliminated from the selection pool due to the pollution associated with running them indoors. Hydraulic hammers were also eliminated because their design involves proprietary valve technology (Menck 1990a and IHC 1988). Vibratory hammers, which are a less common type of pile hammers, were also eliminated from the selection pool. Vibratory hammers (electric and hydraulic) operate in a method that is fundamentally different from all other types as they do not impact the pile head, but instead use imbalanced centrifugal forces to vibrate the pile into position (Goble 1987). A model vibratory hammer would not be representative of typical pile hammers.

Air operated pneumatic hammers are thought to be best suited for laboratory environments. It is desirable that the laboratory pile hammer operates at a high frequency in order to satisfy the requirements of similitude in modeling the hammer-pile-soil system. The frequency of impact of double-acting[1] hammers is higher than that of single-acting hammers. However, the limited head clearances available for laboratory pile driving results in hammer designs with large diameter-to-length ratios that would prohibit the use of a double acting design. The housing of double-acting hammers must be heavier than the uplift force generated in the hammers top chamber, or else the hammer would lift off the pile prior to impact. If the lab hammer were to be made heavy enough to resist lift off, its weight would represent 10–15% of the pile's static capacity at failure. A laboratory double-acting hammer would not be representative of typical production hammers, and would be cumbersome to operate.

An alternative to pneumatic double acting and differential hammers is electromagnetic hammers such as the one developed at The University of Florida by Gills (1988). Electromagnetic hammers may be used in a centrifuge because the hammer's energy is greatly aided by the centrifugal acceleration, and because a short hammer stroke is generally used. For hammers used to drive piles in a pressure chamber, a very large magnetic field would be required to lift a heavy ram through a stroke of several feet. The magnetic field required to lift the ram would result in electrical noise that would prohibit the acquiring of any data electronically.

This process of elimination resulted in a single-acting pneumatic design being selected as the most suitable design for laboratory pile hammers. Pneumatic hammers are typically controlled using mechanical trip valves, which are machined to allow accurate control of airflow in and out of the hammer chamber. Electrical and remote sensing ("radar") systems are, however, better suited for control of laboratory pile hammers because they are easily adjusted to accommodate the wide range of energies and installation conditions which may be required for a laboratory hammer. An electrical system was chosen to control the hammer because it provides simple but precise control of the hammer's energy and operating frequency.

6.3.2 Hammer Geometry

The capacity of the piles tested in the available pressure chamber often exceed 10 tons. Unlike piles tested in a centrifuge, where driving is greatly aided by centrifugal acceleration, the hammer used to drive piles in a pressure chamber must

[1] Applies also to differential and compound hammers.

deliver sufficient energy under a 1-g field to drive the piles to the design depth. In closed laboratory environments, head clearance is at a premium. The clearance available for the model pile hammer used in this study and its guide system was 40 in. The stroke of the laboratory pile hammer must also be restricted such that the hammer velocity does not exceed 9–10 ft/sec in order to minimize the likelihood of damage of the model pile's instrumentation. The limitation on the hammer's length results in several practical considerations, which are discussed next.

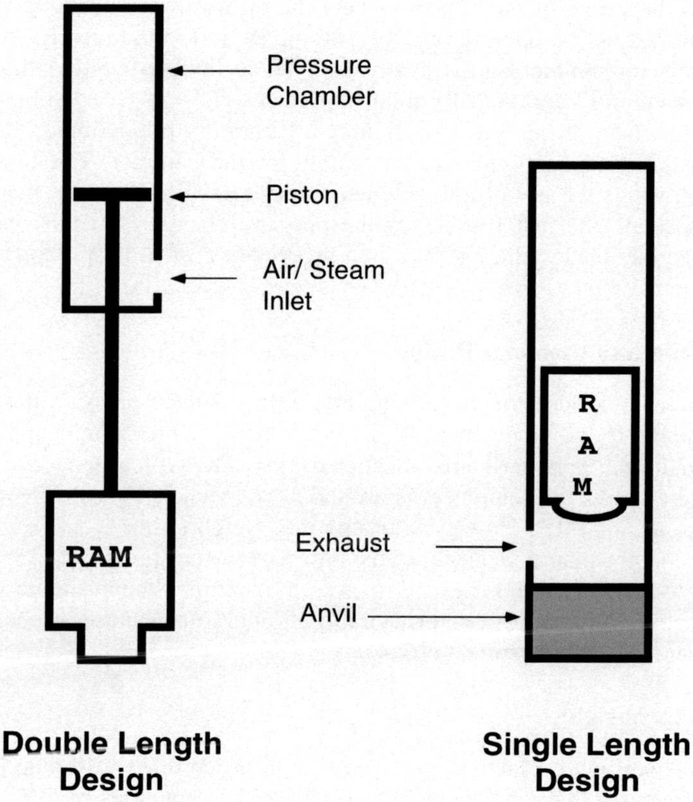

Fig. 6.1 Schematics of Pile Hammers

6.3.2.1 Hammer Shape

The majority of external combustion hammers have a *double length* design, while internal combustion hammers have a *single length* design (Fig. 6.1). In a single length design, the hammer consists of a single chamber in which the ram is lifted by a pressure and dropped to impact the anvil at the bottom of the chamber. Double length hammers consist of a pressure chamber that is located on top of the ram and its guides. The ram is connected to a piston that is located in the pressure chamber. The ram is lifted by introducing a pressure below the piston in the pressure chamber, and is dropped when the pressure is exhausted. The valves that control the flow of air or steam in or out of the chamber are located at the base of the pressure chamber. The advantage of the double length design for external combustion hammers, is that the length of the pressure chamber can be made slightly longer than the stroke of the ram to prevent the piston from impacting the chamber, thus protecting the control valves from impact and vibrations. External combustion air/steam hammers are typically designed as double-length, while internal combustion hammers are typically single length design.

One of the main difficulties in designing a laboratory pile hammer that is used indoors, is the short head clearance available for the hammer. The limited head clearance prevents the use of a double length design, with the result that the control valves cannot be installed below the pressure chamber. In this respect, the shape of the laboratory pile hammer is representative of a diesel hammer rather than a steam hammer.

6.3.2.2 Length to Diameter Ratio

It is important to reduce the cushioning effect of any back pressure that may develop ahead of a ram in a down-stroke, which is achieved in commercial hammers by minimizing the hammer's cross sectional area. The typical length-to-diameter ratio of most production hammers is around 10:1 (Menck 1990b, Pileco-Delmag 1992, and Vulcan 1993). However, for any ram weight, a large-diameter ram occupies less height than a small-diameter ram. Given the limitation in head clearance, a large-diameter ram is needed in order to maximize the available stroke. As a result, the overall length-to-diameter ratio of laboratory hammers is around 4:1, which may reduce the hammer's efficiency.

6.3.2.3 Ram Weight

The piles driven with a laboratory hammer in a pressure or calibration chamber develop capacities that are relatively larger than the capacities of piles driven in soils subjected to a free field effective stress. As a result, the ram weight required is larger than usual. The usual rule of thumb is that the ram weight should be approximately equal to the weight of the pile being driven (Fleming *et al.* 1985). However, wave equation analysis of the driving conditions in a pressure chamber indicates that a ram weighing equal to two to six times the weight of the pile is required.

6.3.2.4 Location of the Control Unit

The control valves of commercial pneumatic hammers are typically mounted below the hammer's pressure chamber in order to reduce exhaust head losses and improve efficiency. However, the use of a single-length design subjects the control valve to excessive vibrations that may damage its electrical components[2]. Accordingly, the hammer and its control unit were designed as two self-contained units, which are connected to one another during driving using flexible pneumatic hoses.

6.4 Layout of the Driving System

The driving system consists of three main components; *the hammer* which contains the ram and other mechanical components, *the hammer leaders* which guide the hammer during pile installation, and *the electro-pneumatic control unit.* The hammer and its leaders are permanently connected (Fig. 6.2). As discussed in Section 6.3.4, the hammer (Fig. 6.3) and its control unit (Fig. 6.4) were designed as two self-contained units that are connected to one another during driving with flexible pneumatic hoses.

6.4.1 The Hammer

The hammer consists of a heavy stainless steel ram that slides vertically in a pressure chamber to impact an anvil at the base of the chamber (Fig. 6.5). The pressure chamber consists of an aluminum cylinder that is capped at the bottom with an aluminum anvil, and at the top with an aluminum top plate. The cylinder, anvil, and the top cap are held together using four stainless steel tie rods. The various components of the hammer were machined to meet close tolerances. The diametrical clearance between the ram and the cylinder is 0.05 inches. The ram is guided in its movement by a pair of ball bushings that ride on a hardened steel rod (Fig 6.5). The rod is aligned in position by having its ends secured inside two holes at the anvil and at the top plate. In an up-stroke, air is introduced at the base of the chamber through holes in the anvil and the ram is lifted upwards. The air supply is closed before the ram reaches its maximum stroke, and the ram continues its upward motion in a *free-ride* under its own inertia. In a down-stroke, the ram free falls under gravitational pull to impact the anvil.

[2] Vibration resistant solenoid valves were not commercially available. Nevertheless, it is possible to custom design a mechanically controlled valve that would withstand vibration.

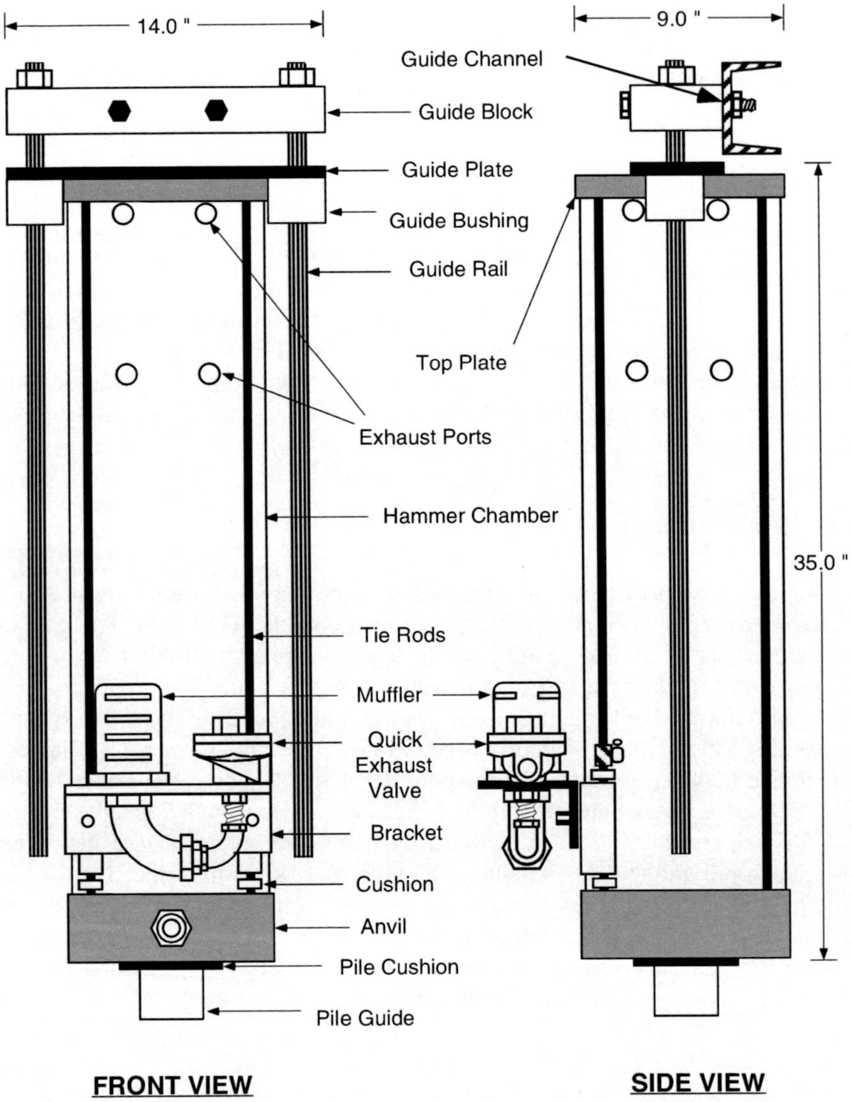

Fig. 6.2 Schematic of the Hammer and Guide System

Fig. 6.3 Photograph of the Pile Hammer and Guide System during Driving

Fig. 6.4 Photograph of the Control Unit

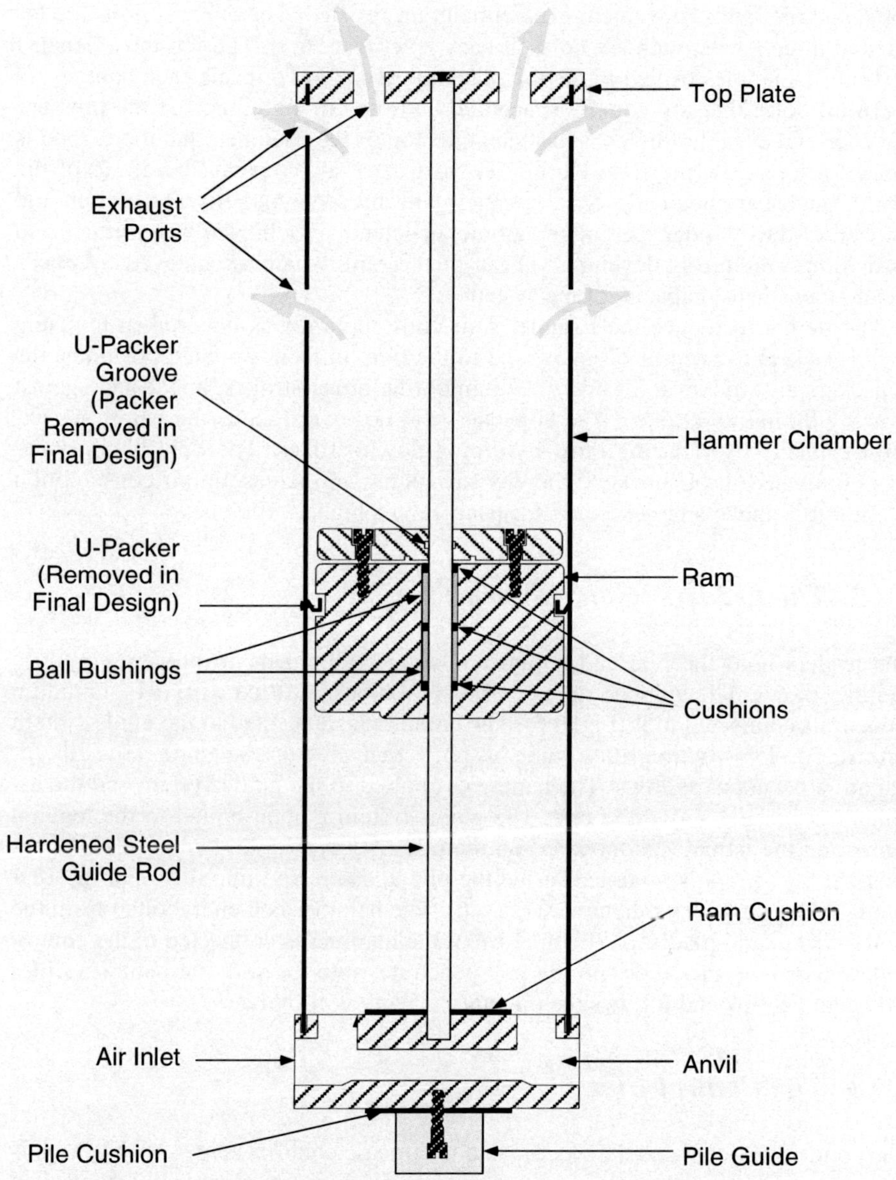

Fig. 6.5 Cross Section of the Pile Hammer

It is important to reduce the amount of air resisting the ram's movement in up-strokes and down-strokes in order to achieve optimum operating speed and efficiency. During up-strokes, air is exhausted ahead of the ram through 18 exhaust ports, and the ram's movement is essentially un-resisted. The exhaust ports are located at three levels, with six holes at each level (Fig. 6.5). The top-most exhaust ports are 1 in. holes drilled in the top plate, the second two levels each contain six 13/16 in. holes that are equally spaced around the circumference of the pressure chamber. One level of holes is located at the top of the chamber, and the second is located 8 in. below the first. During down-strokes, air is exhausted ahead of the ram through the pneumatic control system and by *blow-by* between the ram and the wall of the cylinder. Optimum hammer efficiency is achieved when little or no cushioning pressure is developed ahead of the ram, which is achieved by maximizing the exhaust capacity of the system.

An attempt to reduce the hammer's air consumption was undertaken by using two U-packers to prevent blow-by next to the ram. In their mounted positions, the U-packers allowed air to blow-by the ram during down-strokes, but sealed against blow-by during up-strokes. The U-packers did however reduce the hammer's exhaust capacity by reducing blow-by during down-strokes. The final hammer design eliminated the U-packers, thereby sacrificing an increase in air consumption in order to achieve a larger exhaust capacity and a higher efficiency.

6.4.2 The Leaders (Guide System)

The leaders hold the pile and hammer in proper alignment to prevent eccentric hammer blows. The guide system consists of a block of HDPE with two aluminum guide rails connected to it (Fig. 6.5). The hammer is connected to the guide system through two low-friction HDPE guide bushings that slide on the guide rails. Pile installation occurs as follows: The hammer is bolted to the guide system and the assembly is hoisted above the pile. The guide system is then bolted to the loading frame and the hammer is mounted on the pile. A low-friction Nylon or HDPE cylindrical *pile guide* is mounted inside the pile and screwed into the anvil in order to align the pile with the hammer (Fig 6.6). The hammer is then unbolted from the guide system and placed on the pile. Next, the hammer is connected to the control unit, and driving proceeds. As the pile penetrates into the soil, the hammer slides on the guide rails and follows the pile until driving is complete.

6.4.3 The Control Unit

The control unit contains the electric and pneumatic controls (Fig. 6.4). The ram is powered in its up-stroke by air provided to it from an electrically controlled solenoid valve, which is located at the control unit. The operating frequency and stroke of the hammer are controlled by adjusting the frequency and symmetry on a function generator that initiates the electrical control signal.

6.5 Mechanical Design of the Hammer

Wave equation analysis was used to design the mechanical components of the hammer. The analysis was mainly concerned with determining the weight and size of the ram and anvil, as well as determining the thickness and material of the hammer and pile cushions. Another emphasis in the design was to ensure that the hammer be light enough to facilitate manual set up in a lab, yet sufficiently robust to withstand many hours of pile driving. Wave equation analysis did not, however, provide sufficient precision to determine the length of the force-time impulse generated on impact. Accordingly, a separate analysis was performed to study the force-time signal and determine the *dwell time* of contact between the ram and the anvil.

6.5.1 Wave Equation Analysis

Wave equation analysis is an analytical technique that is used to study the driving properties of a soil-pile system. The analysis is performed by discretizing the pile and the driving system into a number of short segments. Each segment is represented by a mass and a spring, which have properties that approximate those of the segment they represent. The soil resistance is modeled using a series of elastic, perfectly plastic springs, and a linear dash pots. The transition of the soil spring from elastic to perfectly plastic occurs at a deformation that is referred to as the *quake*. Since the mass of the soil is neglected, artificial damping must be introduced to the system to account for the radiation damping, which occurs in the field. The analysis is performed using computer programs such as WEAP (Goble and Rausche 1976), or TTI (Edwards 1967 and Lowery *et al.* 1968). The choice of quakes and damping factors used in the analysis has a significant effect on the computed response of the pile. The computer programs are calibrated against well-documented case histories, which improves the accuracy of the analysis for full-scale production piles. It is, however, not clear that the design parameters which were developed for full-scale production piles, are appropriate for the analysis of model piles driven in a pressure chamber. The damping factors are particularly suspect, since radiation damping in a pressure chamber is likely to be different from radiation damping in the field. Accordingly, for laboratory pile driving, it is also not clear that the measured static capacity of piles is equal to the predicted capacity obtained from wave equation analyses.

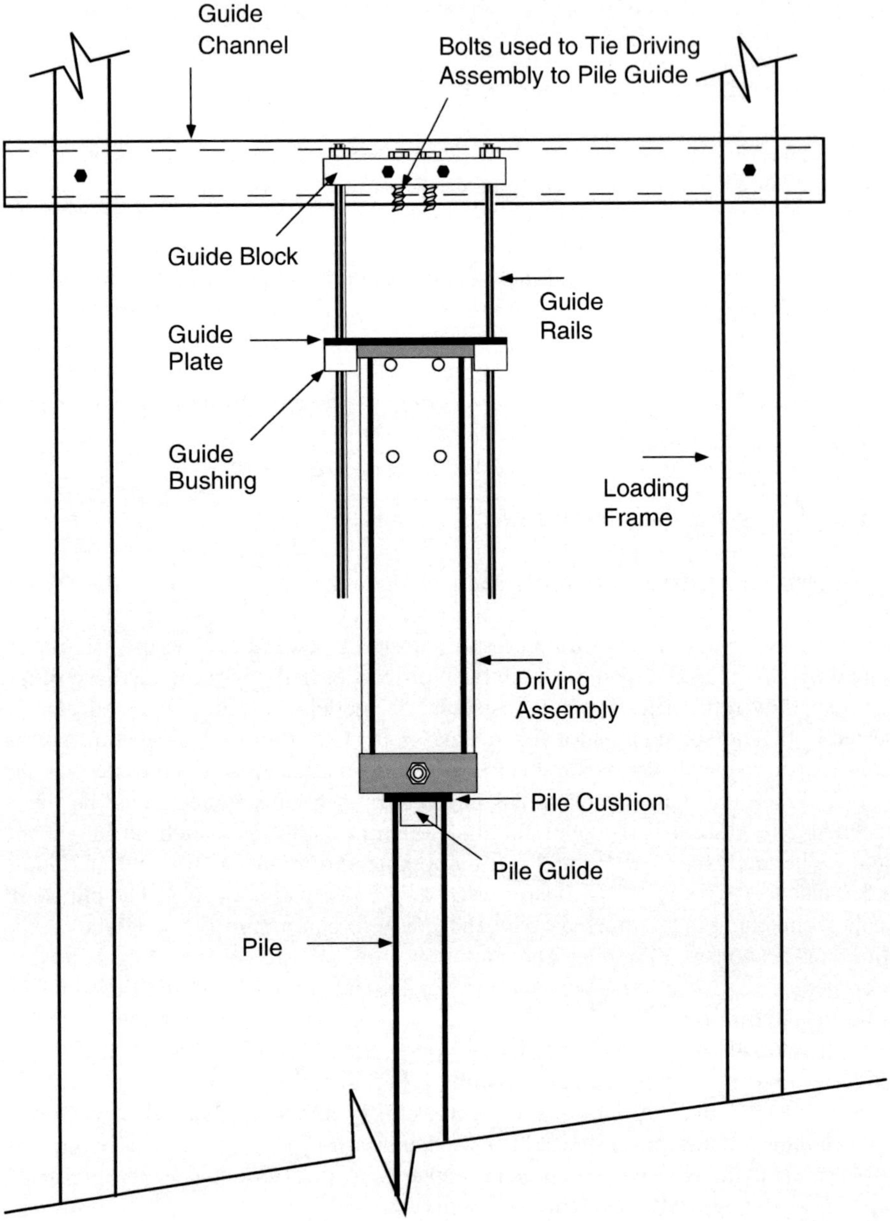

Fig. 6.6 Details of the Guide System

A two-step procedure was used to overcome the difficulty in selecting the proper design parameters for use in wave equation analysis of the hammer-pile-soil system. First, wave equation analysis was performed for the driving conditions of the pile tested by Ugaz (1988) at The University of Houston. The analysis was performed using the computer program WEAP87, and the values of quakes and damping factors that are recommended by GRL (1988) for full-scale piles (Table 6.1). At the final blow count, the capacities predicted using wave equation were found to be 60% of the static capacity measured by Ugaz's (1988) in load tests (Fig 6.7). The ratio of the measured and predicted capacities is assumed to encompass all the errors introduced by the use of the quakes and damping factors intended for production piles. Next, multiple wave equation analyses were performed in order to obtain the optimum dimensions and weights of the mechanical components of pile hammer being designed. The analyses used the wave equation parameters recommended for production piles (Table 6.1). The ratio between the measured and predicted capacities in these analyses was assumed to be the same as that determined for the piles driven in the University of Houston pressure (calibration) chamber (Fig. 6.8).

Multiple analyses of the driving system yielded the hammer specifications shown in Table 6.2. Point A on Fig. 6.8 represents the ultimate static capacity obtained from a load test on a 3.5 in.-diameter, 32 in.-long, closed ended pile which was used in this chapter to evaluate the driving system performance (Fig. 6.9). The design drawings of the ram, anvil, top plate, and guide system are shown in Fig. 6.10.

Table 6.1 Selected Properties Used in Wave Equation Analysis

Side Damping	0.05 sec/ft
Tip Damping	0.15 sec/ft
Side Quake	0.1 in.
Tip Quake	0.05 in.
Hammer Cushion Modulus	175.0 ksi
Pile Cushion Modulus	175.0 ksi
Pile Material Modulus	28500 ksi

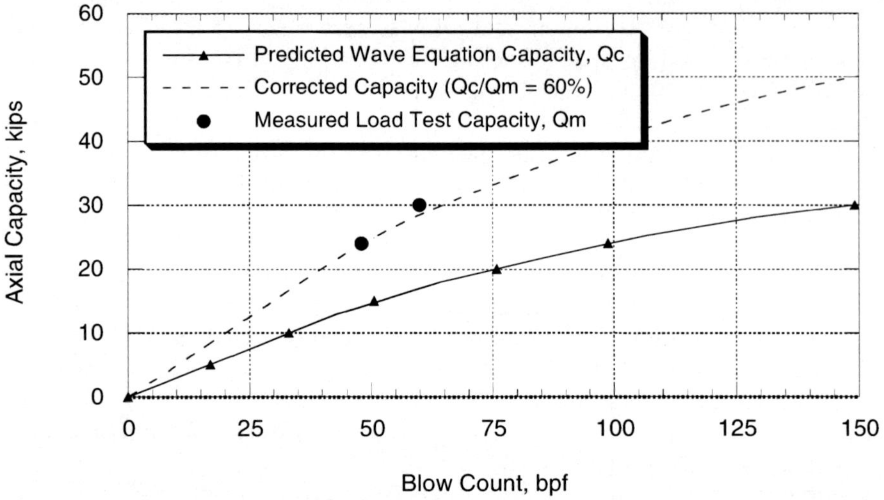

Fig. 6.7 Wave Equation Analyses of Ugaz (1988) Load Tests

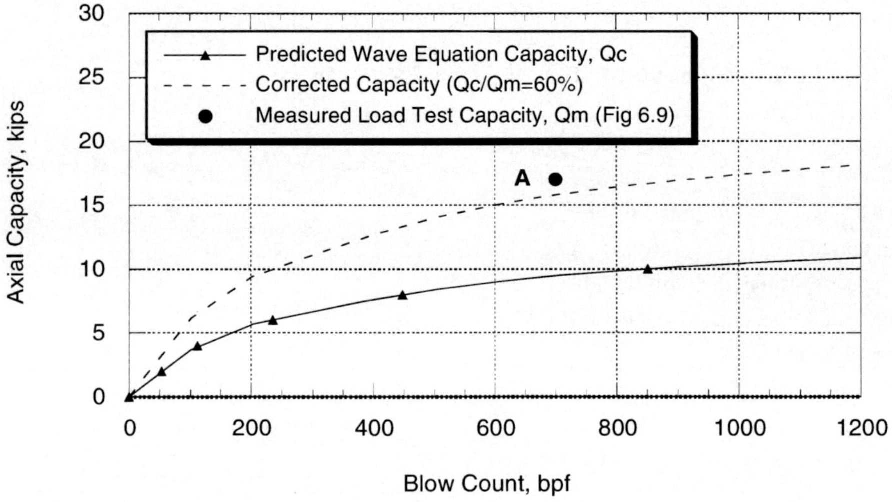

Fig. 6.8 Wave Equation Analyses for Hammer Design

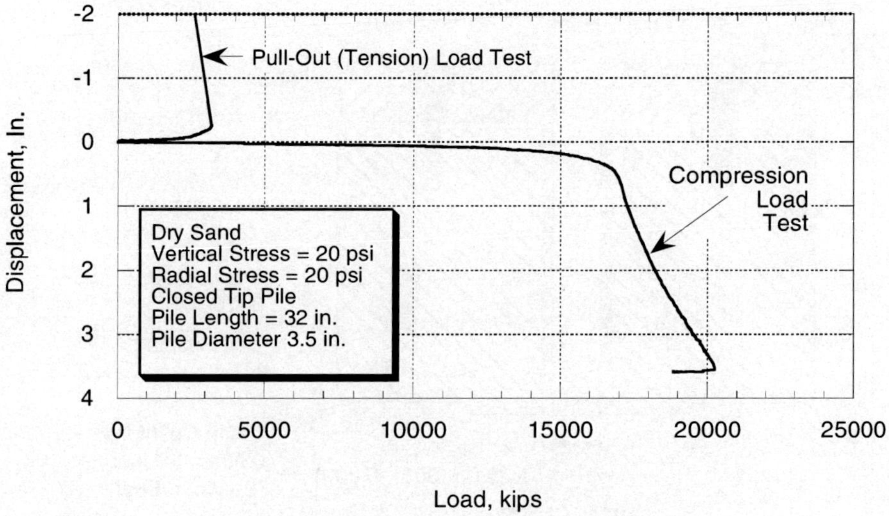

Fig. 6.9 Load Test on Closed-Ended Steel Pipe Pile

Table 6.2 Specifications of the Pile Hammer

Ram Weight	85 lb
Ram Dimensions	8.2 in. in diameter, 6 in. long
Anvil Weight	30 lb
Anvil dimensions	9 in. x 9 in. x 3 in.
Normal Stroke	22 in.
Maximum Stroke	24 in.
Rated Energy	156 lb ft
Hammer Cushion	Nylon, 0.0625 in. thick, 4.5 in. diameter
Pile Cushion	Nylon, 0.0625 in. thick 4.5 in. diameter

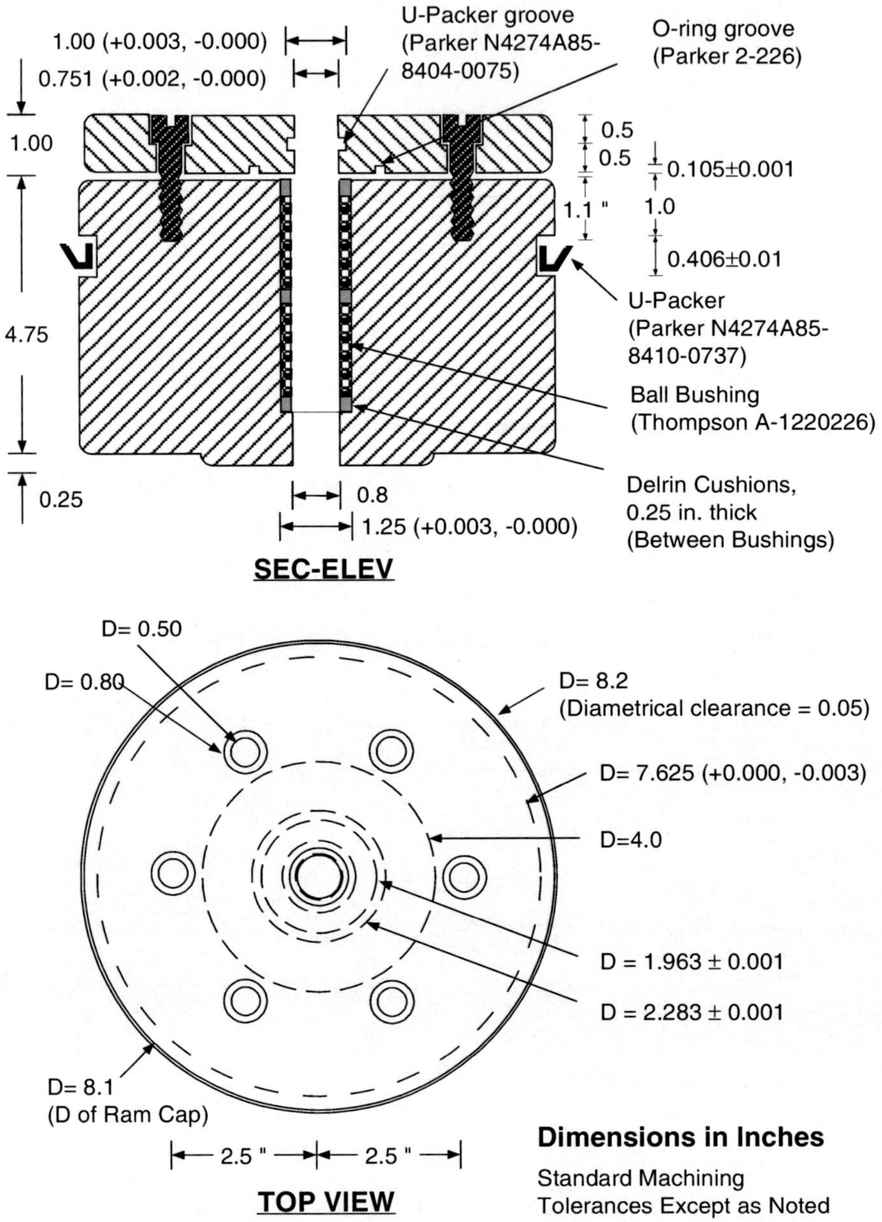

Fig. 6.10a Details of the Hammer: The Ram

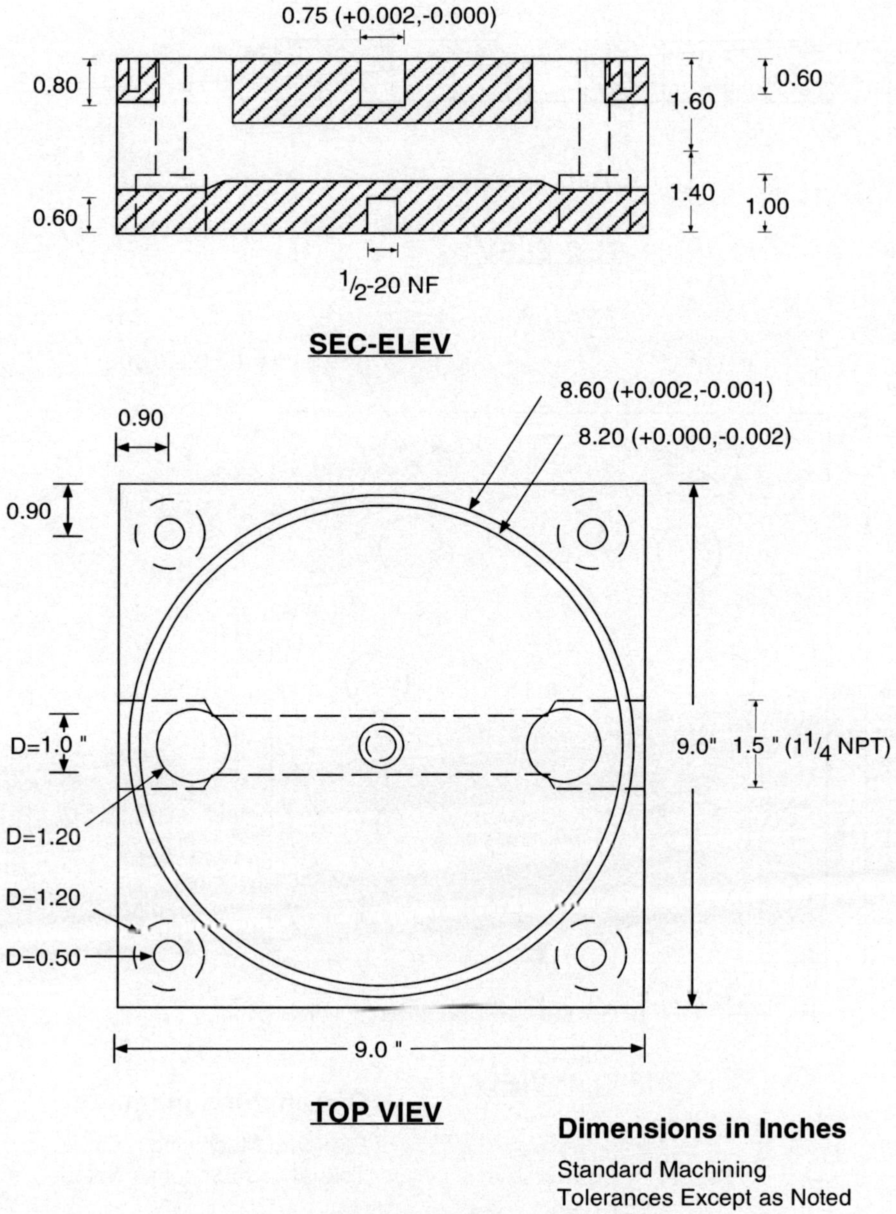

Fig. 6.10b Details of the Hammer: The Anvil

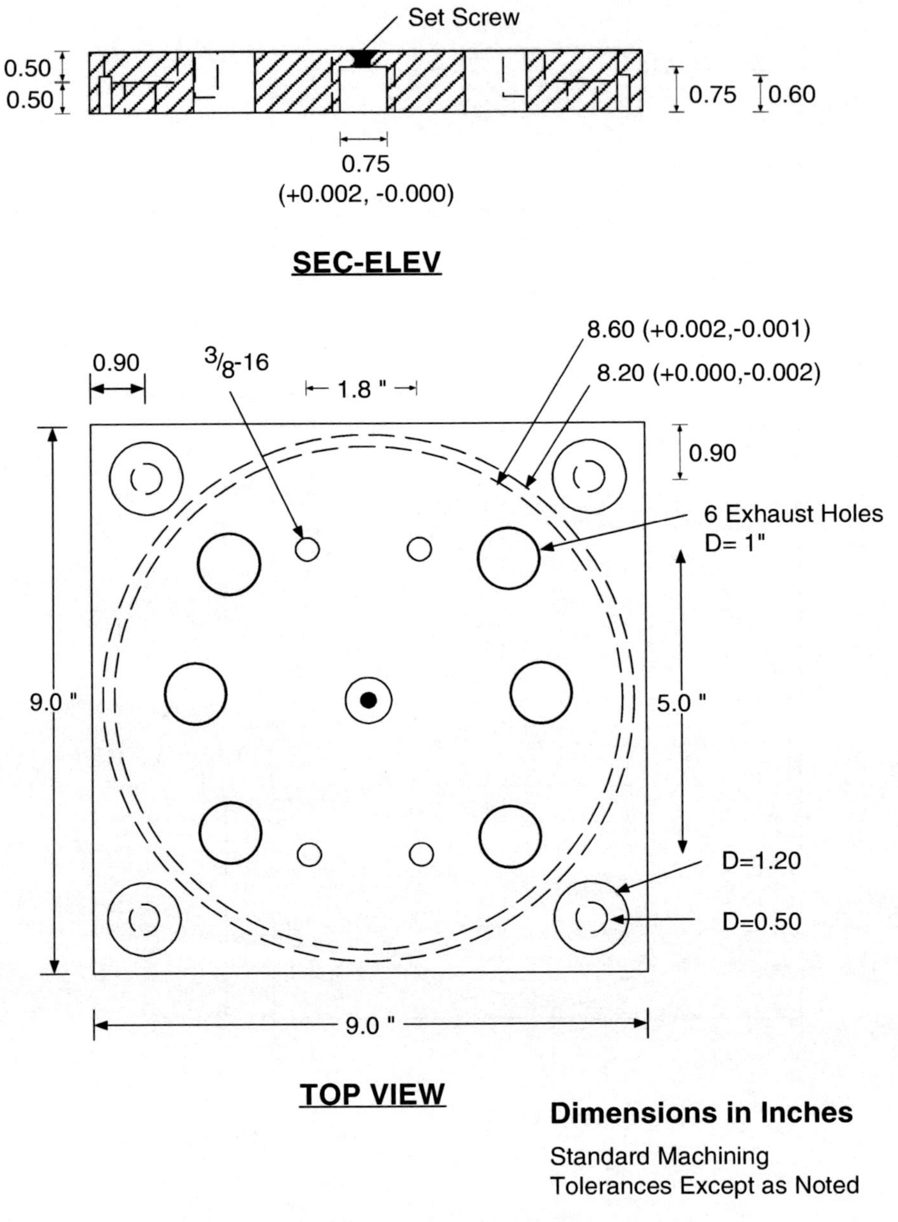

Fig. 6.10c Details of the Hammer: The Top Plate

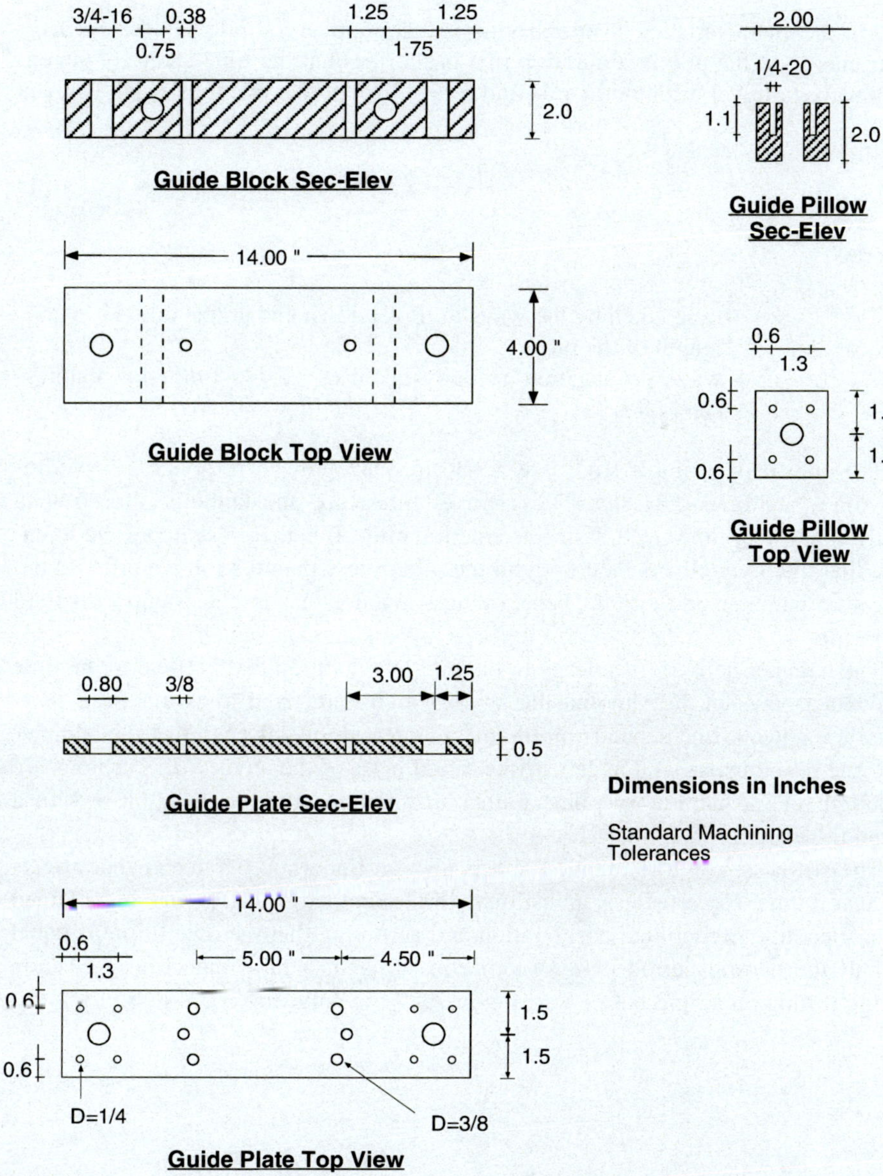

Fig. 6.10d Details of the Hammer: The Guide System

6.5.2 Analysis of the Hammer's Force-Time Signal

One of the characteristics of pile driving is that commercial pile hammers deliver their energy to the pile in a duration that is shorter than the time taken by the energy wave to travel through the pile and reflect back to the pile top, which is given by:

$$t^* = \frac{2 \cdot L}{c} \tag{6.1}$$

where:

t^*	time taken by the wave to travel down and up the pile
L	length of the pile
c	wave propagation velocity taken as 16,400 ft/sec for stainless steel piles

The piles driven in this study are 32–36 in. long. The corresponding wave travel times are 0.33–0.37 ms. These travel times are substantially shorter than equivalent travel times in full-size production piles. It is therefore important to ensure that the ram delivers its energy to the pile in less than 0.33 ms in order to ensure that hammer-pile system behaves in a manner that is consistent with field behavior.

The hammer delivers its energy to the pile in a duration that is equal to the time of contact between the ram and the anvil, which is referred to as the *dwell time*. The ram's dwell time depends on the magnitudes of the stiffnesses of the hammer and pile cushions, as well as the stiffness and mass of the anvil. Stiff cushions are required for the laboratory pile hammer in order to attain a sharp blow with a dwell time shorter than time 2L/c.

The stiffness of the foundation soil is also an important parameter that affects the dwell time. Nevertheless, an estimate of the ram's dwell time may be obtained by neglecting gravitational acceleration and assuming that the dwell time is equal to half the natural period of a system consisting of a ram impacting a cushion spring resting on a rigid base (Warrington 1987), as follows:

$$t_d = \pi \sqrt{\frac{m}{k_e}} \tag{6.2}$$

$$k_e = \frac{k_P \cdot k_R \cdot k_A}{k_P \cdot k_R + k_R \cdot k_A + k_P \cdot k_A} \tag{6.3}$$

$$k = \frac{E \cdot A}{L} \tag{6.4}$$

where:

t_d	dwell time (also referred to as ram period)
m	mass of the ram
k_e	equivalent stiffness of the ram cushion, anvil, and pile cushion
k_P	stiffness of the pile cushion
k_R	stiffness of the ram cushion
k_A	stiffness of the anvil
E	modulus of elasticity (taken as 350 ksi for Nylon and 10,000 ksi for Aluminum)
A	area of cushion or anvil (taken as circle of diameter 4.5 in. for the ram cushion, and as a ring of diameters 4 and 2.5 in. for the pile cushion)
L	thickness of cushion or anvil (taken as 0.0625 in. for the ram and pile cushions)

The ram dwell time is obtained by substituting for the dimensions of the anvil and cushions in Eqs. 6.2–6.4. The ram's dwell time increases with the increase in the thickness of the ram and pile cushions (Fig. 6.11). A theoretical dwell time of 0.3 ms is obtained for the cushion dimensions shown in Table 6.2. The theoretical dwell time is shorter than the time taken by the energy wave to travel through the pile and reflect back to the pile top (Eq. 6.1).

An alternative analytical model for the shape of the force-time record was developed by Deeks and Randolph (1993). The model uses lumped masses to represent the ram and anvil, and a combined spring and dash pot for the cushions. The model was developed for long piles which are represented by a dash pot of impedance $Z=EA/c$, where EA is the cross sectional rigidity as defined above, and c is the wave propagation velocity in the pile. The model's stability is, however, sensitive to the various stiffnesses and masses used. Deeks and Randolph (1993) provided guidance for dealing with imaginary roots in the ranges of stiffnesses and masses of commercial cushions and pile hammers. The cushion stiffness of the model pile hammer is one to two orders of magnitude larger than that of commercial pile hammers. As a result, a mathematical singularity occurred when the model was used to model the behavior of the laboratory pile hammer, and thus the model could not be used.

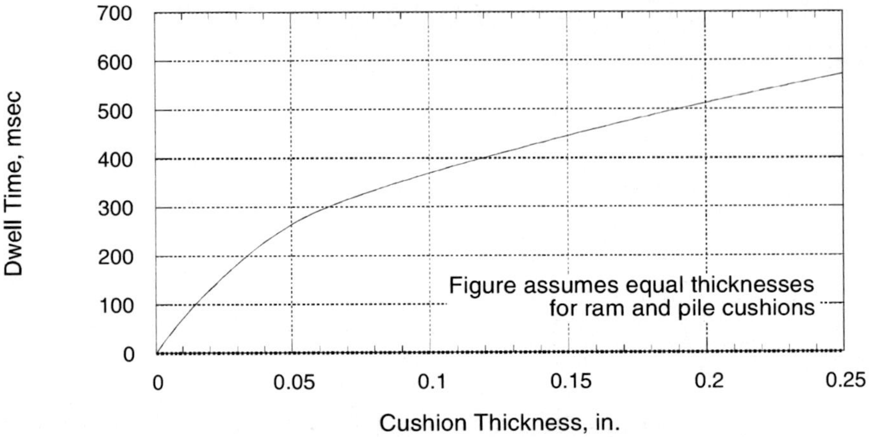

Fig. 6.11 Effect of Cushion Thicknesses on the Ram's Dwell Time

6.6 Design of the Electro-Pneumatic Control System

6.6.1 Basic Principles of Gaseous Fluid Flow

In industry, design methods and procedures involving gaseous fluid flow are almost exclusively based on experimental data (Miller 1984). The principles of compressible fluid flow are rarely used in sizing pneumatic components due to the complexity of the computations. Compressible fluid flow computations involve the principles of thermodynamics as well as fluid mechanics, which gives rise to a plethora of equations. More importantly, the lack of theoretical flow coefficients for the various components of a pneumatic system precludes the possibility of a theoretical analysis of any realistic pneumatic system. Accordingly, empirical methods based on the capacity coefficient are typically used for design.

6.6.1.1 Capacity Coefficient

The capacity coefficient (valve flow coefficient) is a dimensionless index that is used by industrial designers to simplify the problem of control valve sizing. The capacity coefficient is the number of US gallons per minute of water at 60°F which will pass through a given flow restriction with a pressure drop of one psi (Considine 1957)[3]. The flow of any gas or liquid through a restriction with a known pressure drop may be expressed in terms of gallons of water at one psi pressure drop. For gases, the capacity coefficient , c_V, is given by :

$$c_v = \frac{Q}{22.48}\sqrt{\frac{GT}{(P_1 - P_2)P_2}} \tag{6.5}$$

where:

Q	flow in standard cubic feet per minute (at 14.7 PSIA and 60°F)
P_1	inlet absolute pressure in psi (gauge pressure +14.7)
P_2	outlet absolute pressure in psi (gauge pressure +14.7)
G	specific gravity of flowing medium
T	absolute temperature of flowing medium (460 + °F)

For computations involving exhausting a pneumatic chamber, like the hammer chamber, capacity coefficients are usually simplified and expressed as (Parker 1991):

$$c_v = \frac{a \cdot S \cdot A \cdot C}{t \cdot 29} \tag{6.6}$$

where:

a	area of the chamber in square inches
S	stroke in inches
A	constant for pressure drop (from tables in Parker 1991)
C	compression factor (from tables in Parker 1991)
t	stroke time in seconds

[3] The metric equivalent of the capacity coefficient is the flow factor, *Kv*, and is defined in mixed SI units as the number of cubic meters per hour of water at 20°C which will flow with a pressure drop of one bar (kg/cm^2). The flow coefficient, A_V, is a version of the flow factor in coherent SI units. A_V represents the number of cubic meters per second of water at a temperature of 20°C that will flow through the valve with a pressure loss of one Pascal (N/m^2). The three flow factors c_V, K_V, and A_V are interrelated by the following relationships $k_V = 0.85\ c_V$ and $A_V = 2.4 \times 10^{-5}\ c_V$ (Zappe 1987).

To select any component of a pneumatic system, the required capacity coefficient, c_V, is determined first using Eq. 6.6 on the basis of the volume, pressure, and time of flow as well as the acceptable pressure drop across the component being designed. Next, a component is selected from the pool of commercial components such that the capacity coefficient of the selected component is larger than the required value[4].

6.6.1.2 Standard Volume

Gases are compressible and, hence, their volume must be defined at a certain temperature and pressure. Standard volume is the volume of a gas at a standard atmospheric pressure (14.7 psi) and temperature (60°F). Standard volume is useful in performing mass balance of gaseous flow, and is expressed in English units as standard cubic feet (SCF). Flow is expressed in English units as standard cubic feet per minute (SCFM).

6.6.2 Selection of the Pneumatic Components

The minimum capacity coefficient and air consumption required for optimal operation of the hammer are determined next.

6.6.2.1 Required Capacity Coefficient

The most important consideration in sizing the pneumatic components of the hammer is ensuring that the ram falls freely and impacts the anvil with little or no air cushioning. Cushioning of the ram is prevented by ensuring that the air is capable of exhausting from the hammer chamber in less time than is needed for the hammer to fall freely under gravitational pull. The time, t, required by the ram to travel the distance, S, from the bottom of the exhaust ports to the anvil under uniform gravitational acceleration, g, is given by:

$$t = \sqrt{\frac{2 \cdot S}{g}} \tag{6.7}$$

A ram travel (stroke) time of 0.34 sec is obtained by substituting the ram's normal stroke (22 in.) in Eq. 6.7. A corresponding capacity coefficient, c_V, of 29.6

[4] Often only one value of c_V is reported for commercial components. This value represents the average c_V for flows resulting in pressure drops of 2, 5, and 10 psi across the component.

is obtained by substituting the stroke time, hammer dimensions and assuming an inlet pressure, P_1, of 10 psi, and an outlet pressure, P_2, of 8 psi in Eq. 6.6. The 2 psi pressure drop across the pneumatic system was recommended by Parker (1991) for critical applications. A smaller head loss would result in improved hammer efficiency, but would require much larger pneumatic components. Accordingly, all pneumatic components should have a minimum exhaust c_V of 30 in order to prevent formation of a backpressure that cushions the ram's impact.

6.6.2.2 Required Air Consumption

The hammer's air consumption is directly proportional to its operating pressure. The force required to just lift the ram is theoretically equal to the weight of the ram. The operating pressure corresponding to this force is given by:

$$P_T = \frac{W_R}{A_R} \tag{6.8}$$

where:

P_T	theoretical operating pressure required to lift the ram
W_R	ram weight
A_R	cross sectional area of the ram

The theoretical pressure required to lift the ram is obtained by substituting the ram weight and dimensions in Eq. 6.8, and is found to be 1.6 psi. In practice, however, the actual operating pressure required to lift the ram ranges between two to five times the theoretical value (Warrington 1994). The required operating pressure is therefore 3–8 psi[5]. This estimate is dependent to a large extent on the amount of blow-by which occurs between the ram and its housing. If substantial blow-by occurs, the required operating pressure may greatly exceed the estimated value.

The nominal air consumption of the hammer in the range of possible operating pressures is summarized in Table 6.3. The hammer's air consumption and maximum airflow are computed from the operating pressure as follows:

$$V_C = a \cdot S \cdot \left(\frac{P_R + P_a}{P_a} \right) \tag{6.9}$$

[5] The ratio between the actual and theoretical operating pressures represents the uniform design acceleration of the ram in upstroke motion. In practice, however, the ram does not move under uniform acceleration and a uniform design acceleration is of limited value in predicting the hammer's performance.

Table 6.3 Air Usage of the Hammer

Average Operating Pressure	Air Consumption Per Blow	Air Consumption Per Minute @ 1.25Hz	Maximum Air Flow
psi	SCF	SCFM	SCFM
5	0.88	66	160
10	1.11	83	202
15	1.33	100	242
20	1.55	116	282
25	1.78	134	324

$$Q_C = \frac{V_C}{t} \tag{6.10}$$

where:

V_C air consumption per blow in SCF
a hammer chamber cross sectional area
S hammer's stroke
P_R operating pressure in psi
P_a atmospheric pressure in psi
Q_C air flow rate
t time taken by the ram to travel distance S (computed using Eq. 6.3)

6.6.3 The Electro-Pneumatic Circuit

The hammer is powered by compressed ("shop") air, which is obtained from the university utility lines (Fig. 6.12). Compressed air, pressurized at 100 psi, is supplied to the control unit from a 3 in. diameter main through two independent 1/2 in. NPT lines. The flow capacity of each line was measured and found to be approximately 80 SCFM. At the control unit, compressed air is stored in a 60 gallon tank, which functions as a buffer (line tamer) that maintains the pressure of the air supply constant. The tank holds 60 SCF of compressed air and provides a reserve supply of compressed air for the surges of air into the hammer during up-strokes.

At the control unit, compressed air flowing to the hammer passes through a 40 micron filter, followed by a high flow pressure regulator and an electric solenoid valve. The filter (Parker Model # 08F64A) has a flow capacity of about 475 scfm, and produces a pressure drop of about 2 psi at the range of operating flows. The pressure regulator (Parker Model # 08R613A) has a flow capacity of about 400 scfm.

Flow of air in and out of the hammer chamber is controlled using an externally piloted electrical solenoid valve (Parker Model No: N3658900253P). The valve is a 3-way, 3-port, 2-position valve and has an exhaust cv of 26.9 and a fill *cv* of 23.3[6]. The solenoid valve is controlled by a solid-state relay valve (Omega Model No: SSR240DC10), which in turn is controlled by a function generator (Hewlett Packard Model No: HP3312A). The solenoid is energized and opens to allow flow through the valve when it receives AC current from the relay. The relay is a switch that allows AC current to energize the solenoid valve when it receives a positive DC control signal from the function generator. The fill and exhaust times of the hammer chamber are thus controlled simply by adjusting the frequency and symmetry of the control signal generated by the function generator. The electrical system is capable of operating at frequencies up to 120 Hz.

The solenoid valve was selected because it is one of the fastest high capacity valves commercially available. The estimated[7] time taken by the valve to fill or exhaust the volume of air required per blow of the hammer is shown in Fig. 6.13. During the fill cycle, when the solenoid is energized, it moves a short distance to open the pilot gate. The pilot gate controls a high pressure air supply. When the pilot gate is opened, the pilot pressure pushes the solenoid valve's gate all the way and allows air to flow through the solenoid valve into the hammer. During the exhaust cycle, the solenoid is de-energized and the pilot gate closes, which allows air passing through the valve to shut the solenoid valve, thus preventing the flow of air to the hammer and exhausting the hammer chamber to the atmosphere. The pilot mechanism was selected because it provides a very fast and consistent response.

Compressed air that passes through the solenoid valve is divided with a tee into two separate lines. The first line is a 1-1/4 in. NPT hose that connects directly to the hammer chamber. The second line is a 3/4 in. NPT hose that connects to the hammer chamber through a quick exhaust valve (Parker Model #OR75B50) which is located on the driving assembly. The quick exhaust valve allows air to flow from the controller to the hammer chamber, but exhausts the air directly to the atmosphere during hammer down-strokes. The quick exhaust valve increases the exhaust capacity of the system by reducing the distance traveled by the air to exhaust, thus permitting greater hammer speeds without increasing the size of the controlling solenoid valve. The quick exhaust valve, which was the largest commercially available valve in the United States at the time, has a fill c_v of 8.5, and an exhaust c_v of 11.1 (Ellenger 1993).

---

[6] Exhaust and fill air pass through different orifices in the solenoid valve which results in different exhaust and fill capacity coefficients.

[7] Computed from data provided courtesy of Ellenger (1993).

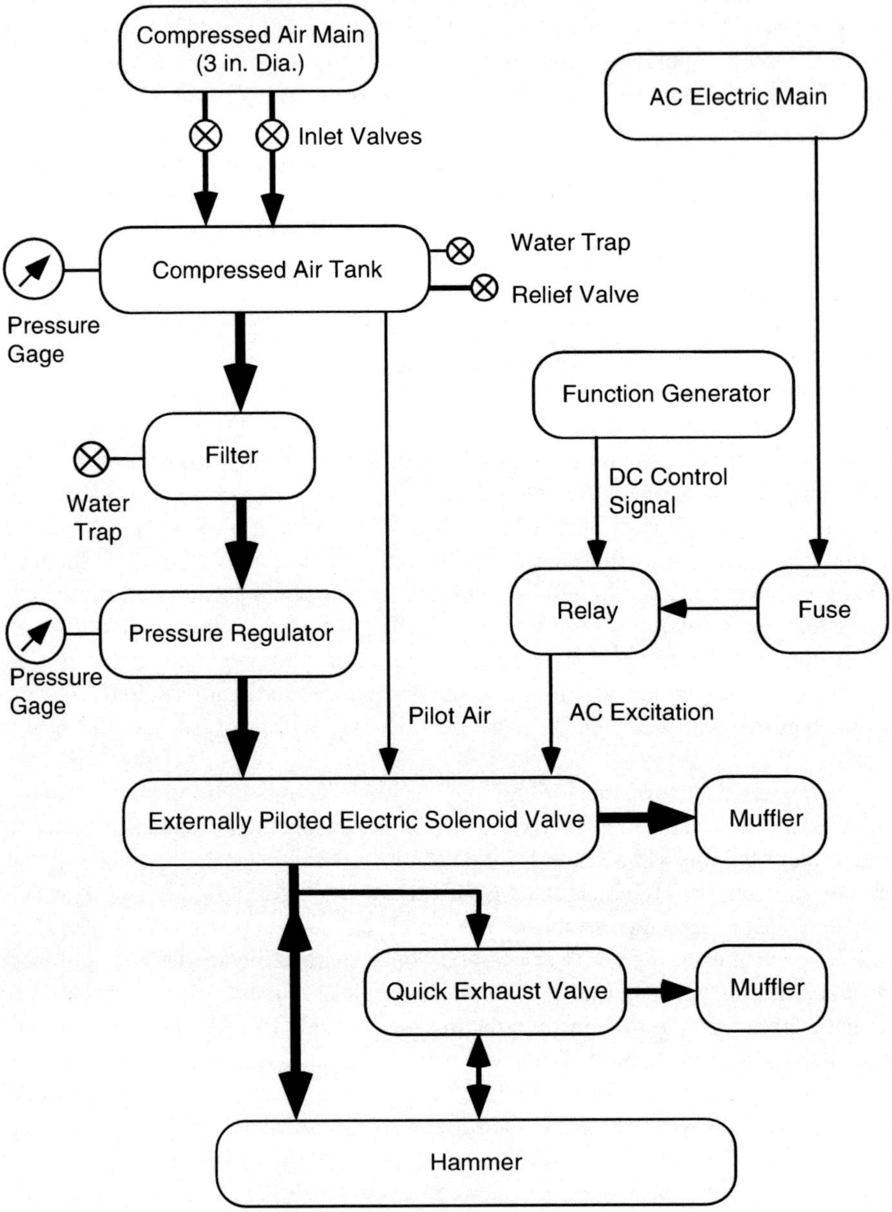

Fig. 6.12 The Electro-Pneumatic Circuit of the Pile Hammer

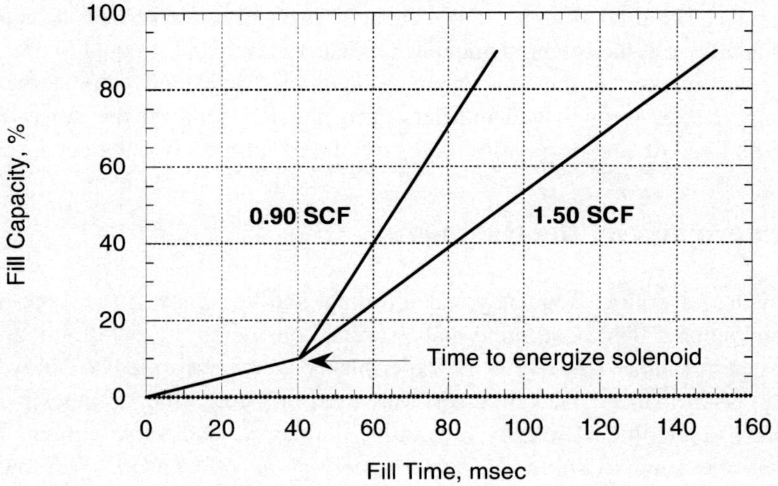

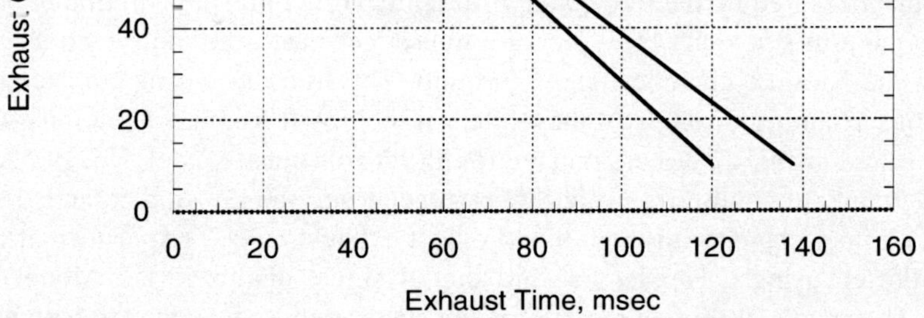

Fig. 6.13 Performance of the Solenoid Valve during Fill (Top) and Exhaust (Bottom)

The air stream was divided into two lines because the Parker OR75B50 quick exhaust valve, which is a 3/4 in. valve, was considered too small to exhaust all the air efficiently to the atmosphere. It is not possible to synchronize multiple quick exhaust valves in parallel, and accordingly only one valve was used to provide secondary exhaust capacity in addition to the main exhaust capacity through the solenoid valve. The total system exhaust capacity is equal to the sum of the capacity coefficients, c_v, at the solenoid and quick exhaust valves and is equal to 38.

Exhaust air, released to the atmosphere at the quick exhaust valve and at the solenoid valve passes through two mufflers. The mufflers reduced the noise level from 115 db to 92 db and have a flow capacity of 470 scfm each (Ellenger 1993).

6.7 Performance of the Hammer

The efficiency, operating frequency, valve timing, and the shape of the force-time record are among the most important criteria considered in determining the hammer's performance. Two sets of experiments were performed to study the hammer's performance. The first set was used to study the *hammer's raw performance*, including efficiency, operating frequency, and valve timing. The second set was used to study the *performance of the hammer-pile-soil system during driving*.

6.7.1 Raw Performance of the Hammer

Raw performance experiments were performed with the hammer placed on a 1 in. Neoprene cushion, which allowed prolonged and repetitive testing of the hammer under identical conditions. The hammer was instrumented with a 30 in. LVDT (TRANSTECK Model No: 223) and a piezoelectric pressure transducer (Omega model No: PX181). The LVDT was threaded to the ram, and extended out of the hammer through one of the exhaust ports in the hammer's top plate. The displacements measured by the LVDT were differentiated with respect to time in order to obtain the ram's velocity. The pressure transducer measured the pressures generated in the hammer chamber below the ram. The transducer was connected to the hammer chamber through a hole drilled in the anvil. Ram accelerations were also measured using an accelerometer (PCB Piezotronics Model No: 305A04) mounted on an aluminum rod, which was threaded into the ram and extended outside the hammer chamber through the top-most exhaust holes. Experimental data were collected using a Hewlet Packard digital signal analyzer (HP Model No: 3562A). The results of this set of experiments were used to evaluate the hammer's efficiency and repeatability, as well as to evaluate the performance of the electro-pneumatic control system.

6.7.1.1 Efficiency

The potential energy of a single-acting pile hammer represents the maximum theoretical (rated) energy of the hammer. During the hammer down-stroke, potential energy is converted to kinetic energy, such that the energy delivered to the pile on impact depends only on the mass of the ram and its impact velocity. If all the hammer's potential energy were to be converted to kinetic energy the ram would impact the anvil with a theoretical velocity, V_t, given by:

$$V_t = \sqrt{2gh} \tag{6.11}$$

where:

V_t	theoretical ram impact velocity
g	gravitational acceleration
h	normal hammer stroke

The actual energy of a single-acting hammer is less than the theoretical energy due to friction losses and the energy consumed in exhausting the air through the pneumatic system. The hammer's efficiency is the ratio between the actual energy delivered by the hammer to its potential energy. The efficiency of the hammer, η, can be obtain by comparing the actual impact velocity of the ram, V_i, with the theoretical impact velocity as follows:

$$\eta = \frac{V_i^2}{V_t^2} = \frac{V_i^2}{2gh} \tag{6.12}$$

where:

η	hammer efficiency
V_i	actual ram impact velocity

The pressures, displacements and velocities obtained during preliminary testing of the hammer are shown in Fig. 6.14. Ram lift-off, maximum stroke and impact are represented on Fig. 6.14, by points A, B, and D, respectively. During down-strokes, a small back pressure (2.8 psi) formed below the ram which resulted in the hammer reaching a terminal velocity of 5 ft/sec after a 6 in. stroke (point C). The backpressure corresponds to the head loss required by the pneumatic system to exhaust the air through the solenoid and quick exhaust valves. Small exhaust head losses resulted in a significant reduction in the hammer's efficiency due to the hammer's large diameter-to-length ratio. During preliminary testing, the efficiency corresponding to the measured ram impact velocity was 20%. The maximum operating frequency during the preliminary tests was 60 bpm (1 Hz).

A pressure drop of 2 psi was employed in Eq. 6.6 to obtain the required c_v, per Parker (1991) recommendations. In retrospect the design pressure drop was too high to achieve a higher hammer efficiency.

As a result of the preliminary testing, the hammer design was modified by removing the U-packers from the ram. The U-packers were initially installed to reduce the amount of blow-by, thus reducing the air consumption. Removal of the U packers increased the available exhaust capacity due to blow-by but also increased the air consumption by an estimated 200%. The results of the final performance tests are shown in Fig. 6.15[8]. Points A, B, and C on Fig. 6.15 represent the ram's lift-off, maximum stroke and impact, respectively. The ram reached a maximum velocity of 8.2 ft/sec, which corresponds to an efficiency of 57%. The modification resulted in an increase in the operating frequency to 72 bpm (1.2 Hz).

The measured efficiency of the laboratory pile hammer compares favorably with the efficiency of production pile hammer which were found to range between 25 and 60%. (Chen *et al.* 1979 and Vines and Amar 1979). The efficiency of the laboratory hammer is, however, less than the values reported by hammer manufacturers for use in wave equation analysis (GRL 1988). The difference may result from methods used to obtain the ram's velocity for efficiency computations. In this study, the ram's velocity was obtained directly from the displacement record of the ram, which is considered to be a superior method compared to indirect velocity measurements using radar devices, which are typically used by hammer manufacturers. However, the efficiencies used in wave equation analysis need not bear a direct numerical correlation to actual field efficiencies due to the empirical nature of wave equation parameters.

The energy lost due to build up of a pressure below the ram, during impact, may be obtained by integrating the work done in cushioning a blow as follows:

$$E_c = \int_0^{Stroke} p_s \cdot a \cdot s \cdot ds \tag{6.13}$$

where:

E_C	energy lost due to cushioning
s	distance traveled by the ram
P_S	pressure below the ram at given ram position
a	hammer's cross sectional area

[8] The ringing in the DC control signal results from the use of an anti-aliasing filter in the HP 3562A analyzer. During polarity reversal, the control signals contain energy in the filtered-out high frequency range which appears as ringing (Gibbs Phenomenon). The pressure and displacement records did not contain energy at the filtered-out frequencies and their records were actually cleaned by the use of a filter.

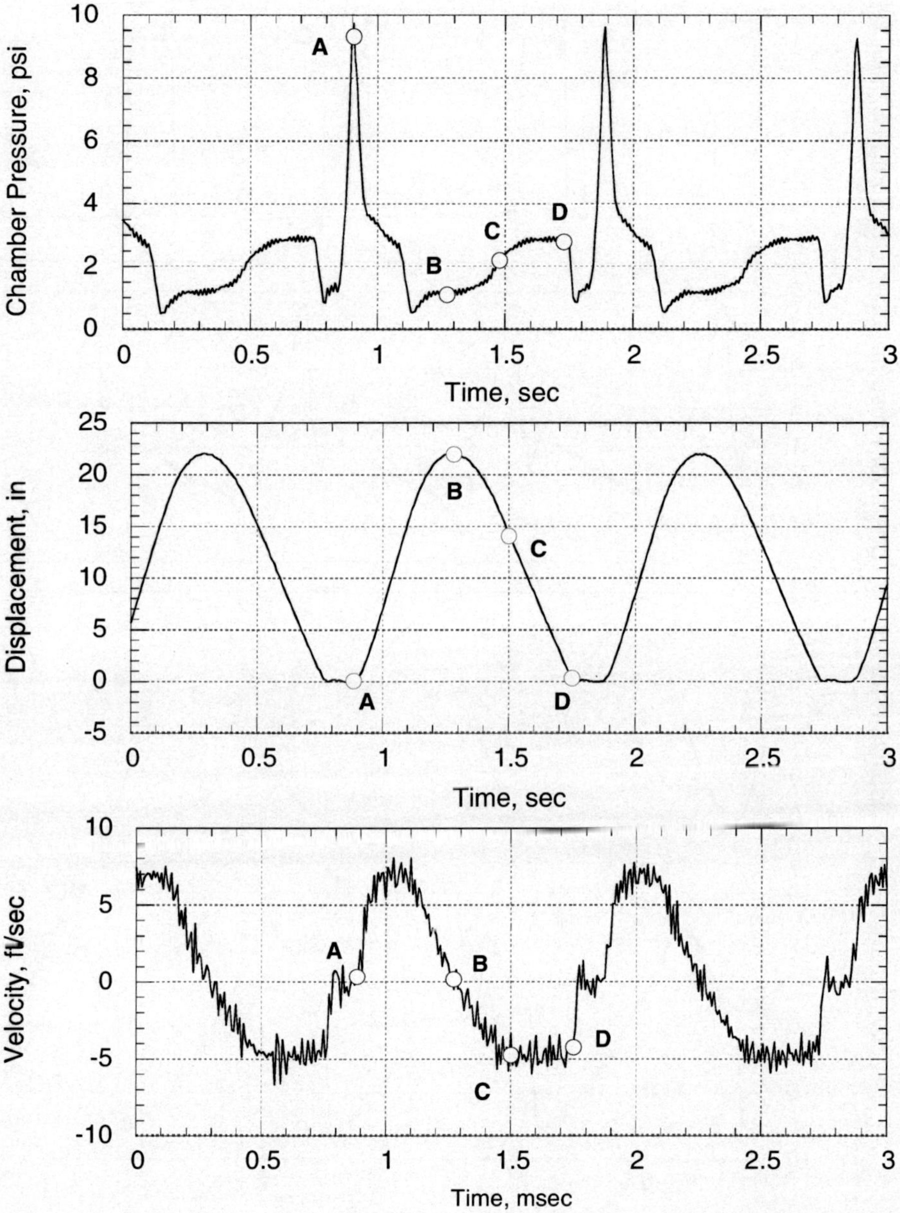

Fig. 6.14 Preliminary Hammer Performance (Before Modification)

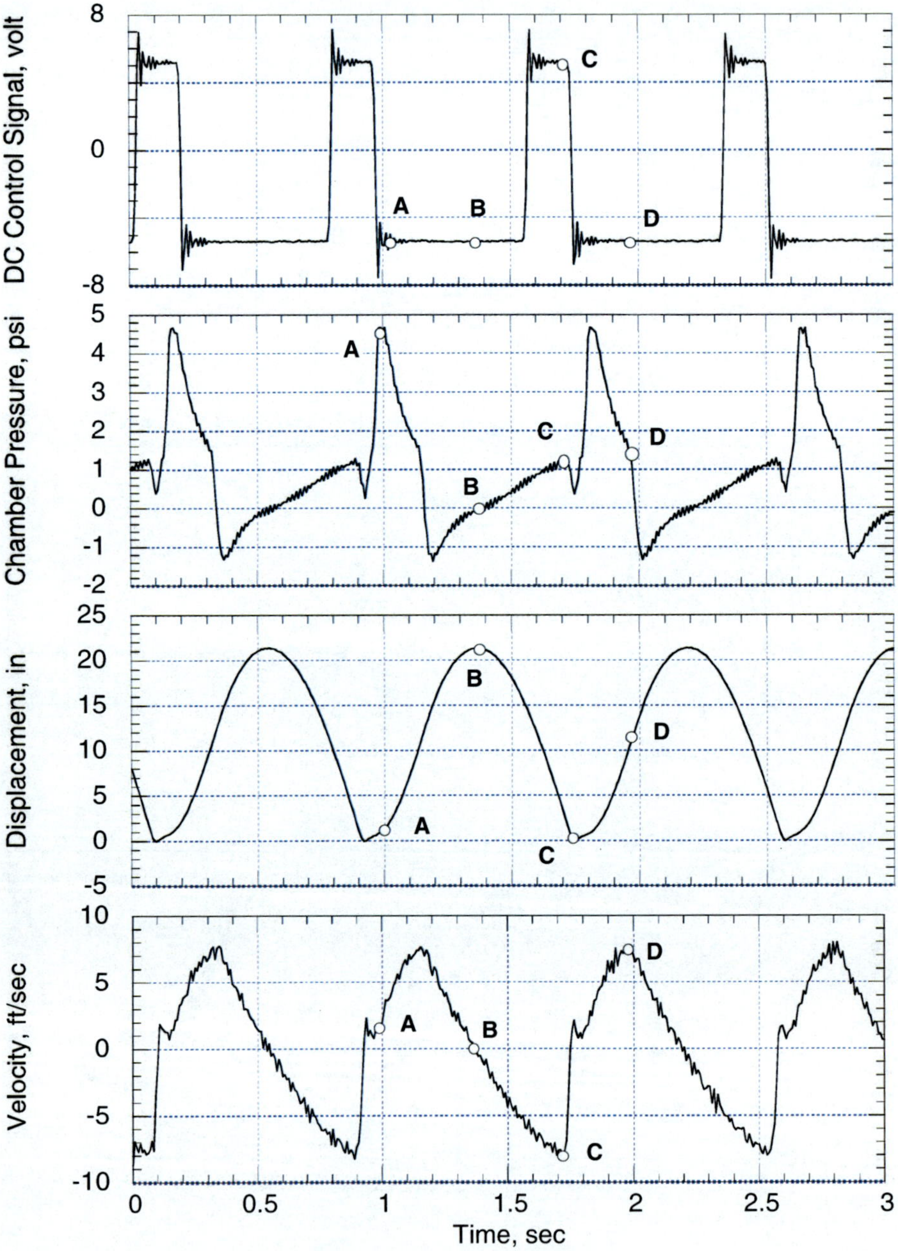

Fig. 6.15 Final Hammer Performance

Equation 6.13 was used to determine the energy loss due to cushioning for a number of successive hammer blows. The computation indicates that 43–45% of the ram's potential energy is lost due to cushioning. Therefore, considering that the measured hammer efficiency is about 57%, it is believed that friction results in negligible energy losses.

6.7.1.2 Performance of the Electro-Pneumatic Control System

The function generator, relay, and solenoid valve provided excellent control of the air into the hammer chamber. It was possible to control the airflow such that the ram is energized in up-stroke right after it rebounds from impacting the anvil (Point A on Fig. 6.15). The airflow was terminated before the hammer reached full stroke (Point D on Fig 6.15), such that the ram exercised a *free ride* under its own inertia, thereby reducing the air consumption and the volume of exhaust air.

Air control during exhaust left much to be desired. The head losses through the pneumatic components appear to be larger than those reported by the manufacturer. The difference may be attributed to the fact that commercial pneumatic components are designed and tested for continuous flow, and suffer during intermittent flow. The quick exhaust valve provided marginal benefit because most of the air was exhausted either by blow-by or at the solenoid valve, perhaps due to the slow frequency response of the quick exhaust valve.

Better control of the exhaust may be achieved if the distance traveled by the exhaust air is shortened, which could be done by mounting the solenoid valve directly below the hammer chamber. This was not possible due to the use of a single length design (Section 6.3.4). The solenoid valve used is not sufficiently robust to withstand the shock waves generated at the anvil.

6.7.1.3 Repeatability

The hammer performed with outstanding repeatability, which may be judged by comparing ram impact velocities over a number of successive blows. The mean ram impact velocity and standard deviation over the twenty successive blows shown in Fig. 6.16 were 8.2 ft/sec and 0.19 ft/sec, respectively.

Hammer repeatability during pile driving depends on soil conditions. The ram rebound does not depend on the hammer cushioning, but rather on the soil's resistance, particularly for short piles (Warrington 1994). As pile driving progresses and the pile develops additional resistance, the ram rebound velocity increases, and less air is required to lift the ram in an up-stroke. The ram velocities reported in this study were measured with the hammer placed on a Neoprene cushion which provided limited resistance to driving. As a result, the measured ram rebound represents a lower bound value. Accordingly, the measured efficiency of 57% is also considered a lower bound value.

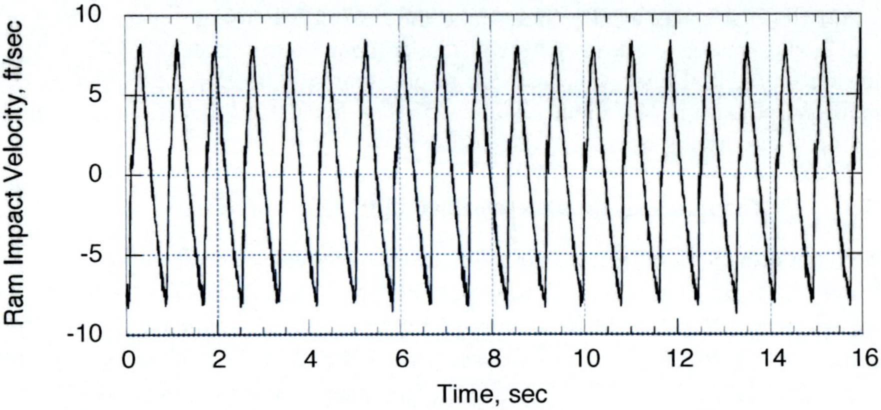

Fig. 6.16 Ram Impact Velocity during Successive Hammer Blows

6.7.1.4 Ram Acceleration

The ram impacted the anvil with an acceleration reaching 2,000 G and rebounded with accelerations reaching 1,000 G (Fig. 6.17). The velocity of the ram at impact was obtained by integrating the acceleration record and was found to be 8.5 ft/sec (Fig 6.18).

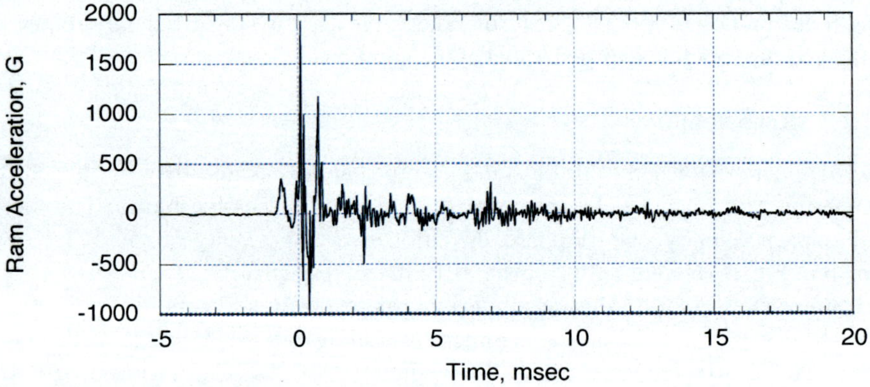

Fig. 6.17 Ram Acceleration During Impact

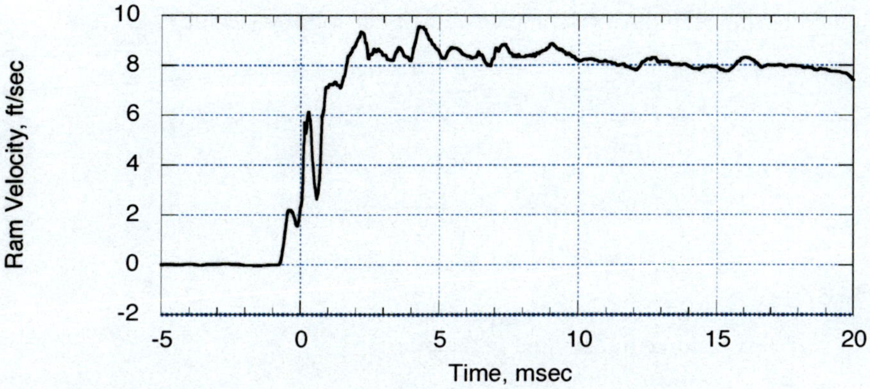

Fig. 6.18 Ram Velocity during Impact

6.8.2 Hammer Performance during Pile Driving

The hammer was used to drive a number of piles in the pressure chamber shown in Chapter 4. The piles were 3.5 in. in diameter, 33 in. long, and had a 0.12 in. wall thickness. The pile head was instrumented with an accelerometer (PCB Piezotronics Model No: 3405A04), a 0.5 in. DCDT (TRANSTECK model No: 243), and an axial strain gage, which were mounted 0.8, 1.9, and 4.5 in. below the pile top, respectively. The instruments measured accelerations, displacements and forces, respectively, during each blow. The measured accelerations were integrated with respect to time to obtain the pile head velocities. Experimental data was collected using a Hewlet Packard digital signal analyzer (HP model No: 3562A). The results of this set of experiments was used to study the behavior of the hammer-pile-soil system.

6.8.2.1 Force-Time Curve

The total energy delivered by the hammer is not the sole criterion for effective pile driving. The shape of the force-time curve plays an important role in determining the efficiency of pile driving. A curve with a high peak force and short duration may have the same energy as a curve with a lower force over a longer time, but the prolonged force-time curve is more effective in driving piles (Fuller 1983).

The transmission of the forces from the hammer to the pile can be expressed as being proportional to the ram's impact velocity as follows (Goble *et al.* 1980):

$$F = \sigma \cdot A = \frac{E \cdot A}{c} V_i \tag{6.14}$$

$$c = \sqrt{\frac{E}{\rho}} = \sqrt{\frac{E \cdot g}{\gamma}} \tag{6.15}$$

where:

F	force transmitted to the pile
σ	stress transmitted to the pile
E	modulus of elasticity of the pile taken as 28,500 ksi for stainless steel
V_i	impact or particle velocity
c	celerity or wave propagation velocity in the pile taken as 16,400 ft/sec for stainless steel
ρ	mass density of the pile material
γ	unit weight of pile material
g	gravitational acceleration

The pile head forces and velocities averaged over 200 blows are shown in Fig. 6.19. The outstanding overlap of force and velocity records up to a length of *2L/c* is indicative of the good quality of the data. The force-velocity-time record may be explained as follows. On impact, a compressive wave propagates through the pile and is reflected at the pile tip as a tension wave. The force and velocity records are identical up to a duration equal to *2L/c*, when the reflected wave reaches the pile top. The hammer, which is much heavier than the pile, provides relative fixity to the pile head and reflects the tension wave downward as a magnified tension wave. The tension wave cancels compression waves, and a mismatch of force and velocity is evident at the second velocity peak, which occurs after a duration equal to *2L/c* from the first peak. The tension wave is reflected at the pile tip as a compressive wave, which is in turn reflected at the hammer as a magnified compressive wave which peaks at a duration equal to *4L/c* after the first peak.

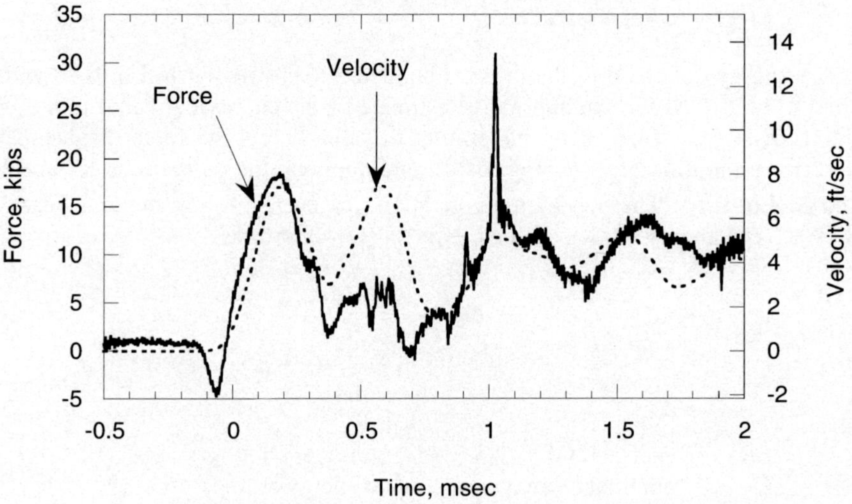

Fig. 6.19 Pile Head Force and Velocity Records

6.8.2.2 Energy Delivered to the Pile

6.8.2.2.1 Energy Delivered According to ASTM D4633

The energy delivered to the pile, which is different from the hammer's energy, is given by ASTM (1993) Standard Test Method for Stress Wave Energy Measurements for Dynamic Penetrometer Testing Systems (D4633-86) as:

$$E_i = \frac{c \cdot K_1 \cdot K_2 \cdot K_c}{A \cdot E} \int_0^{\Delta t} F^2(t) \;\; dt \tag{6.16}$$

where:

E_i	energy in the first compression pulse for an infinitely long pile
c	wave propagation velocity
A	cross sectional area of the pile
E	modulus of elasticity

K_1	correction factor to account for the energy not recorded by the strain gages, taken as 1.15
K_2	correction factor for short piles, taken as 1.9
K_c	correction factor for the wave propagation velocity, taken as 1
$F(t)$	compressive force

The energy delivered to the pile is obtained by substituting in Eq. 6.16 and is found to be 61.2 ft lb. An alternative estimate of the energy delivered to the pile of 55.7 ft-lb was obtained by integrating the area below the force displacement curve for a duration of 2L/c (Fig. 6.20), and applying the corrections K_1 and K_2 used in Eq. 6.16. These energies correspond to energy ratios ranging between 35–38%. The energy ratio is defined in ASTM (D4633-86) as:

$$ER_i = \frac{E_i}{E^*} \tag{6.17}$$

where:

ER_i	energy ratio
E^*	nominal (Newtonian) kinetic energy at impact
E_i	energy content of the first compression pulse

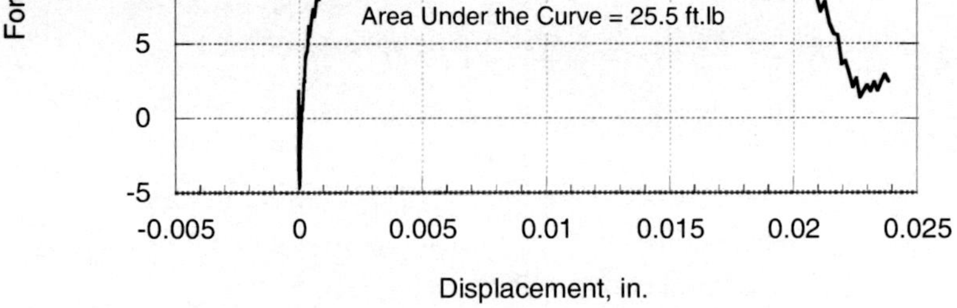

Fig. 6.20 Force Displacement Curve of the Pile

The measured energy ratios are somewhat smaller than the values reported for Standard Penetration Test (SPT) drop hammers which typically range between 55–60% (Seed *et al.* 1984). The computations for the energy ratio depend to a large extent on the corrections used to account for violating the infinitely long pile assumption of the ASTM standard method. ASTM standards provide the values of the corrections for the typical dimensions of the SPT. The corrections used in this study were obtained by extrapolation of the ASTM's data, which may account for the reduction in energy ratios.

6.8.2.2.2 Realistic Estimate of Energy Delivered to Pile

The energy transferred into a rod in an axial impact, *Ei*, is given by Timoshenko and Goodier (1970):

$$E_i = \int_0^\infty F(t)\ v(t)\ dt \tag{6.18}$$

where:

$F(t)$ force measured at a point near the edge of the rod
$v(t)$ velocity measured at a point near the edge of the rod

Butler *et al.* (1998) and Farrar (1998) applied this technique to calculate the efficiency of SPT hammers. The integrated energy input to the pile is calculated in Fig. 6.21 and found to be 103 kip-ft.

This energy corresponds to a hammer efficiency of 66%. And is believed to be a better estimate of the energy delivered to the pile during driving.

6.8.2.3 Ram Dwell Time

The ram's dwell time is considered equal to the duration of the first compression pulse on the force–time record. The force–time record of the energy delivered to the pile (Fig. 6.22) indicates that the duration of the first compression pulse is 0.35 msec, which is equal to *2L/c*. An alternative measure of the dwell time is shown in Fig 6.21d; it was approximately 4 msec. Both estimates are close and are approximately one order of magnitude larger than that calculated using Eq. 6.1–6.3 for the ram impacting the anvil. The system response was dominated by the soft response of the pile-soil system. The dynamic stiffness of the soil-pile system varies within three orders of magnitude during impact (Fig. 6.23). An approximate dynamic stiffness, K_d, can be calculated as follows:

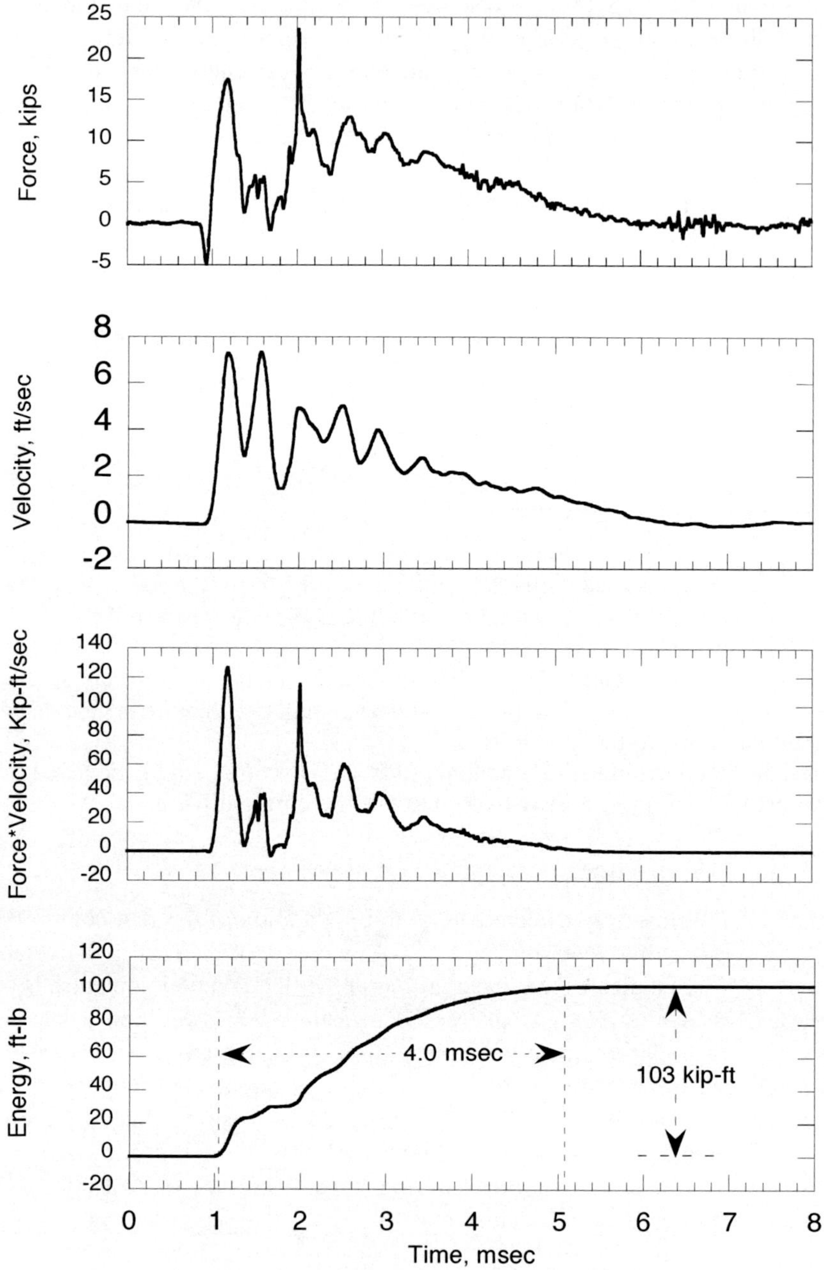

Fig. 6.21 Energy Delivered to Pile. (a) Measured Force, (b) Velocity (Integrated from Acceleration), (c) Force*Velocity, and (d) Energy (Integrated from (c))

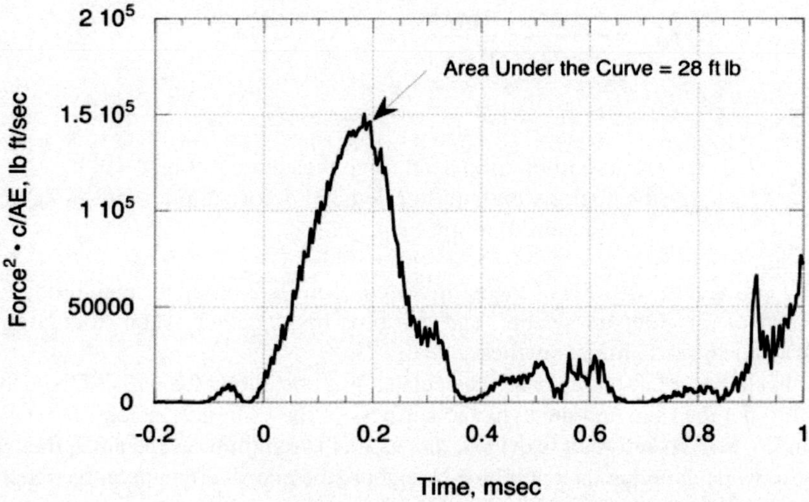

Fig. 6.22 Force–Time Record of Energy Delivered to the Pile

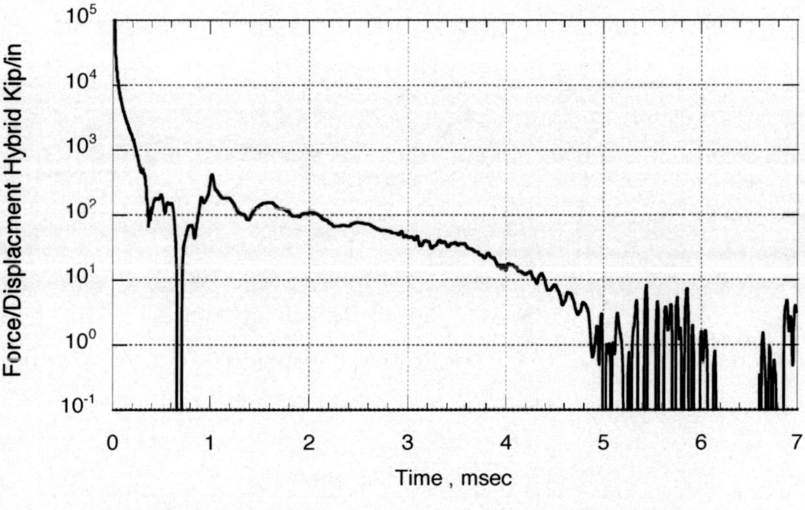

Fig. 6.23 Dynamic Pile-Soil Stiffness during Impact

$$K_d = \frac{F_{\max}(t)}{S_{\max}(t)} = \frac{18\ kips}{0.08\ in.} = 225\ kips/in. \tag{6.19}$$

where:

$F_{max}(t)$ the maximum pile head force measured in Fig. 6.19
$S_{max}(t)$ the average maximum pile head deformation per blow (not final set) measured in Fig. 6.24.

The equivalent stiffness of the hammer/cushions is 2 orders of magnitude larger than that of the pile/soil system, and can thus be neglected when calculating the soil/pile/cushions/hammer stiffness.

A dwell time of 4.0 msec can be calculated by substituting the mass of the hammer (145 lb), not the ram, and the dynamic stiffness of the soil/pile/cushions/hammer system in Eq. 6.1. Therefore, it is evident that when the ram impacts the anvil, they move in phase while applying their combined weight to the pile. The length of the dwell time depends totally on the stiffness of the soil/pile system.

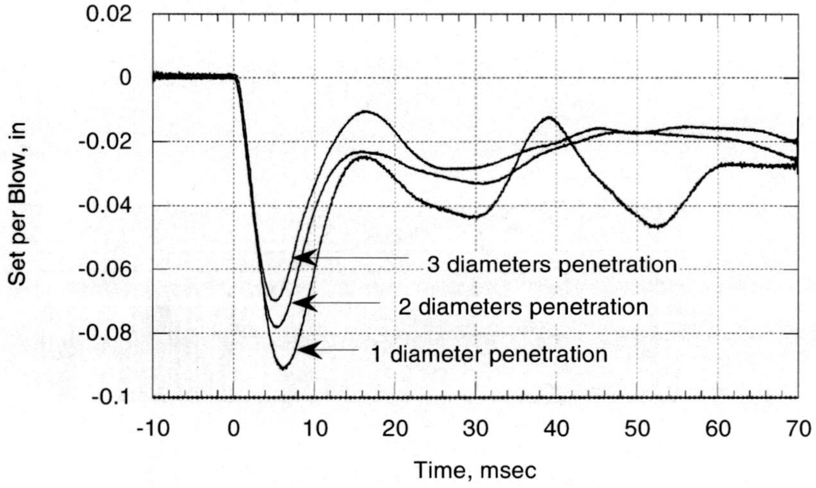

Fig. 6.24 Pile Head Deformation per Blow for a 3.5 in.-Diameter Closed-Ended Pile

6.8.2.4 Pile Set

Upon each blow, the pile head undergoes both elastic and plastic deformations (Fig. 6.24). The plastic deformation that represents the pile set is on the order of 0.02 in., which is in agreement with the results of wave equation analysis (Fig 6.8). The magnitude of the elastic deformation exceeds the elastic compression of the pile material by several orders of magnitude. The measured elastic deformation is therefore representative of the elastic deformation of the soil, particularly in end bearing.

6.8.2.5 Wave Equation Analysis

As expected in the preliminary analysis, WEAP underestimated capacity by about 40%. Piles were driven to a resistance of about 1970 bpm (600 bpf) and the corresponding wave equation capacity was approximately 45 kN (10 kips). Static loading tests yielded a capacity of about 63 kN (14 kips).

The properties used in the initial analyses did not lead to a good match between measured and calculated stresses and pile head velocities. The pile driving process led to shock waves that emanated out to reflect off the sides of the metal pressure chamber. The assumption of a pile in a half space was invalidated at an early stage and consideration was concentrated on the first *2L/c* time period (about 0.4 msec). The best fit was achieved using side and tip quakes of 0.02 inch (the same as the set), side and tip damping factors of 0.4 sec/ft, and reducing the cushions' stiffness by half.

6.9 Conclusions

The methods developed in this study to design a laboratory pile hammer proved to be successful. The hammer delivers a repeatable rated energy of 156 ft lb with a 57% efficiency. The maximum impact velocity and operating frequency are 8.2 ft/sec and 1.2 Hz, respectively. There are a number of lessons that have been learned during the design process:

- The most difficult challenge in the design has been the limited head clearance. Future designers should make every effort to maximize the available head clearance.
- For piles driven in a pressure chamber, the capacity estimated using wave equation appears to represent 60% of the ultimate static capacity measured in load tests.
- The electro-pneumatic control system provided excellent control during upstrokes, and adequate control during down-strokes.

- The capacity coefficient, c_V, is a useful tool in designing pneumatic components. Use of a smaller head loss across the system in Eq. 6.6 may result in better performance.
- Air blow-by the ram provides a dramatic increase in the exhaust capacity of the system, thereby increasing the hammer efficiency. Blow-by increases air consumption and should be allowed as long as sufficient air is available.
- Most of the energy loss occurred due to the development of a cushioning back pressure ahead of the ram, despite the fact that every effort was made to reduce the back pressure during down-strokes. Future hammer designs should attempt to mount the control valve on the hammer, while isolating it from shock and vibration. Also the recommended 2 psi head drop for determining the exhaust c_v value of pneumatic components results in excessive cushioning of the ram in its down stroke.
- Use of a mechanical quick exhaust valve provided marginal benefit due to the large head loss required to exhaust air through it.

References

ASTM, Annual Book of Standards, Vol. 4.08, ASTM 1916 Race St., Philadelphia, PA 19103 (1993)

Allard, M.A.: Soil Stress Field Around driven Piles, Dissertation, California Institute of Technology (1990)

Bernardes, G.D.: Dynamic and Static Testing of Large Model Piles in Sand, Ph.D. Thesis, Norwegian Institute of Technology (NTH), University of Trondheim (1989)

Butler, J., Caliendo, J., Goble, G.: Comparison of SPT Energy Measurement Methods. In: Robertson, P., Mayne, P. (eds.) Proc., Geotechnical Site Characterization, Balkema, pp. 901–905 (1998)

Chen, D.D.S., Toto, J.V., Wong, I.H.: Field Evaluation of Hammer Efficiency and Pile Driving Criteria. In: Fuller, F.M. (ed.) Proc. ASCE Piling Symposium, Atlanta, GA (1979)

Craig, W.H.: Modeling of Pile Installation in Centrifuge Experiments. In: Proc. 11th Int. Conf. on Soil Mechanics and Foundation Engineerting, vol. 2, pp. 1101–1104 (1985)

Considine, D.M.: Process Instruments and Controls Handbook. McGraw-Hill Book Company (1957)

Deeks, A.J., Randolph, M.F.: Analytical Modeling of Hammer Impact for Pile Driving. Int. Journal for Numerical and Analytical Methods in Geomechanics 17, 279–302 (1993)

Edwards, T.C.: Piling Analysis Wave Equation Computer Program Utilization Manual, Texas Transportation Institute, Research Project Report 33-11, Texas A&M University, p. 40 (1967)

Ellenger, W.(Bill): Proprietary Test Data, Parker Hannifin Corporation, Richland, Michigan 49083 (1993)

Farrar, J.: Summary of Standard Penetration Test Energy Measurements Experience. In: Robertson, P., Mayne, P. (eds.) Proc., Geotechnical Site Characterization, Balkema, pp. 919–926 (1998)

Fleming, W.G.K., Weltman, A.J., Randolph, M.F., Elson, W.K.: Piling Engineering. John Wiley and Sons, NY (1985)

Fuller, F.M.: Engineering of Pile Installations. McGraw-Hill Book Company (1983)

Gills, J.J.: Development and Testing of a Device Capable of Placing Model Piles by Driving and Pushing in the Centrifuge, Dissertation, University of Florida (1988)

Goble, G.G., Rausche, F.: Wave Equation Analysis of Pile Driving - weap Program, Report, Prepared for the U.S. Department of Transportation, Federal Highway Administration, Implementation Division, Office of Research and Development (1976)

Goble, G.G., Rausche, F., Likens, G.: The Analysis of Pile Driving. In: A State of the Art, Int. Seminar on The Application of Stress Wave Theory on Piles, Stockholm, pp. 131–161. Balkema (1980)

Goble, G.G.: A Pile Inspector's Guide to Hammers and Pile Driving Systems. Pennsylvania Department of Transportation (1987)

GRL, Documentation of weap 1987 Computer Program, Vol. 1-4 (1988)

IHC, IHC Hydro Hammer, PO Box 26, 2960 AA Kinderdijk, The Netherlands (1988)

Iskander, M., Olson, R., Bay, J.: Design and Performance of an Electro-Pneumatic Pile Hammer for Laboratory Applications. Geotechnical Testing J., GTJODJ 24(1), 72–82 (2000)

Lowery, L.L., Edwards, L.C., Hirsch, T.J.: Use of The Wave Equation to Predict Soil Resistance on a Pile During Driving, Texas Transportation Institute, Research Report 33-10 (1968)

Mehle, J.S.: Centrifuge Modeling of Pile Driving, Master Thesis, Department of Civil, Environmental, and Architectural Engineering, University of Colorado, Boulder Colorado (1989)

Menck GMBH, MHU, Offshore Hydraulic Pile Driving Hammers, Werner-Von-Siemens-Strasse 2, D2086 Ellerau, Germany (1990a)

Menck GMBH, MRBS, Offshore Steam/Air Pile Driving Hammer, Werner-Von-Siemens-Strasse 2, D2086 Ellerau, Germany (1990b)

Miller, D.S.: Compressible Internal Flow. British Hydromechanics Research Association (BHRA) Fluid Engineering Series, BHRA, vol. 10 (1984)

Murff, D.: Personal Communications. Exxon Production Research Company, Houston, Texas (1992)

Nunez, I.L., Hoadley, P.J., Randolph, M.F., Hulett, J.M.: Driving and Tension Loading of Piles in Sand on a Centrifuge. In: Proc. Centrifuge 1988, Balkema, pp. 353–362 (1988)

Parker, Parker Pneumatic Products Catalog 0107-1, Pneumatic Division, Parker Hannifin Corporation, Richland, Michigan 49083 (1991)

Pileco-Delmag, Corporate Brochures and Hammer Data Sheets, PO Box 16099, Houston, TX 77222 (1992)

Pelletier, J.H., Murff, J.D., Young, A.C.: Historical Development and Assessment of the Current API Design Methods for Axially Loaded Piles. In: Proc. Offshore Technology Conf., Houston, TX, pp. 253–282, OTC paper 7157 (1993)

Seed, H.B., Tokimatsu, K., Harder, L.F., Chung, R.M.: The Influence of SPT Procedures in Soil Liquefaction Resistance Evaluations, Earthquake Engineering Research Center Report No: UCB/EERC-84/15, College of Engineering, University of California, Berkeley (1984)

Timoshenko, S., Goodier, J.: Theory of Elasticity. McGraw Hill, New York (1970)

Ugaz, O.G.: An Experimental and Numerical Study of Impact Driving of Open-Ended Pipe Piles in Dense Saturated Sand, Dissertation, University of Houston (1988)

Vesic, A.S.: A Study of Bearing Capacity of Deep Foundations, Final Report on Project B-189, Georgia Institute of Technology, Atlanta, Georgia (1967)

Vines, W.R., Amar, J.: Stress Measurements for Offshore Pile Driving. In: Fuller, F.M. (ed.) Proc. ASCE Piling Symposium, Atlanta, GA (1979)

Vulcan Iron Works, Bulletin 65K-1, Specifications for Offshore Pile Hammers, PO Box 5402, Chattanooga, TN 37406 (1993)

Warrington, D.C.: Personal Communications, Vice Chairman, Vulcan Iron Works, Inc., PO Box 5402, Chattanooga, TN (1994)

Warrington, D.C.: A Proposal for a Simplified Model for the Determination of Dynamic Loads and Stresses During Pile Driving. In: Proc. 19th Offshore Technology Conf., OTC paper No: 5395, Houston, Texas (1987)

Zappe, R.W.: Valve Selection Handbook, 2nd edn. Gulf Publishing Company, P.O. Box 2608, Houston, Texas (1987)

7. Geotechnical Properties of the Testing Sand

Abstract. The sand used in model studies must meet a number of scaling requirements. Several types of sand were selected for screening, and one type was chosen for testing. This chapter presents the factors considered in the selection process, along with the geotechnical properties of the chosen sand.

7.1 Introduction

The soil used in this study has to meet a number of requirements. First, it must be finer than offshore sands in order to meet the scaling requirements of model testing (Bolton and Lau 1988). Second, its particle shape must be representative of sands found in the offshore in order to simulate their behavior, particularly with respect to arching and liquefaction or cyclic mobility under cyclic loading (Holubec and D'Appolonia 1973). Third, it must be commercially available in large quantities and with consistent properties.

This chapter presents the factors considered in the selection process, along with the geotechnical properties of the chosen sand.

7.2 Preliminary Choices

A number of sands were screened for this study. The screening was limited to commercially processed sands because of their general availability, consistency in gradation, and assured quality. The following sands were considered:

- *Oklahoma Sand:* Oklahoma sand is mined from the Simpson Formation, primarily for sand blasting applications. Oklahoma sand is white, rounded, and poorly graded (Fig 7.1). It is similar to Ottawa Sand in many respects. The finest commercial grade of Oklahoma sand was considered (Grade 1).
- *Texblast:* Texblast is mined primarily for sand blasting applications. Texblast is tan, sub-angular to sub-rounded, and poorly graded (Fig 7.1). The finest grade of Texblast sand was considered (Grade 1).
- *SIL-CO-SIL 250:* SIL-CO-SIL is a ground silica powder that is manufactured by U.S. Silica (formerly Ottawa Silica Company and Pennsylvania Glass Sand Corporation), primarily as a filler material for paints, adhesives, and ceramics. It is bright white, angular to sub-angular, and very well-graded (Fig 7.1). Because of its fine gradation, ground silica has a low permeability, on the order of

M. Iskander: Behavior of Pipe Piles in Sand, SSGG, pp. 177–186.
springerlink.com

10^{-6} cm/sec, which is desirable in modeling liquefaction and cyclic mobility effects.

- *Ground Silica Mixture:* A mixture of 70% Texblast and 30% SIL-CO-SIL was also considered for the study. The mixture was easier to handle than pure ground silica but possessed the same low permeability characteristics of ground silica.

The properties of Oklahoma sand, Texblast, and SIL-CO-SIL 250 are summarized in Table 7.1.

7.3 Selection of the Test Sand

Ground silica and ground silica mixtures were eliminated from consideration because silica powder contains free (air-born) silica dust that was difficult to contain in the research area and presented a health hazard if breathed, and also threatened machinery. Additionally, handling and pluviating ground silica was substantially more cumbersome than fine sands.

Oklahoma sand was chosen because it possessed rounded particles and was more uniformly graded than Texblast. Rounded and uniform sands have a number of advantages over angular sands. First, rounded sands are more representative of the sands found in the offshore environment being modeled. Second, when pluviated, uniform and rounded sands form more uniform deposits than well-graded and angular sands. Third, rounded particles are less likely to crush during driving, which would maintain the gradation and frictional properties. Fourth, rounded sands are more susceptible to liquefaction than angular sands (Ishibashi 1985).

Table 7.1 Properties of Sands Under Consideration

	Texblast	SIL-CO-SIL	OK Sand
Unified Soil Classification	SP	SW	SP
Mean Particle Diameter	0.19 mm	9.5 μm[‡]	0.17 mm
Specific Gravity	2.65	2.65	2.65
Max. Dry Density, pcf	114	99	113
Min. Dry Density, pcf	91.8	63	97
Max. Void Ratio	0.8	1.62	0.70
Min. Void Ratio	0.45	0.67	0.46
Coefficient of Curvature, C_c	0.91	1.18	1.15
Coefficient of Uniformity, C_u	2.0	18.3	1.5
Friction Angle (Direct Shear), ϕ	43°	45°	41°
Stainless-Sand Interface Friction, ϕ	25°	—	27°

[‡] Median Particle Size = 45 μm

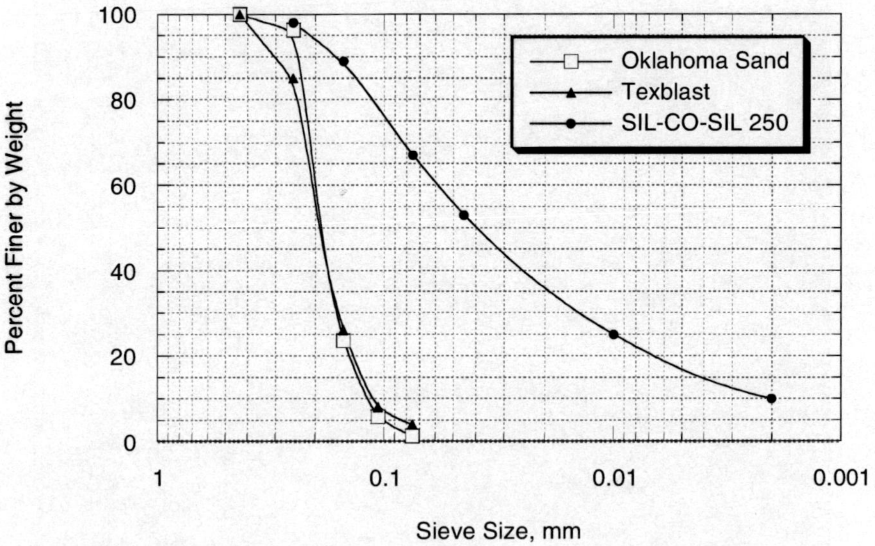

Fig. 7.1 Grain Size Distribution of the Three Selected Sands

7.4 Geotechnical Properties of Oklahoma Sand

Oklahoma sand was delivered in 100 lb bags. The sand was recycled and new sand was added as needed. The geotechnical properties were determined for specimens of new sand.

7.4.1 Physical Properties

Oklahoma sand is rounded to sub-rounded (Fig. 7.2). Grain size analysis was performed in accordance with ASTM (1993) standard D422-63 (Fig 7.1). The sand is poorly graded with a uniformity coefficient, C_u, of 1.5 and a coefficient of curvature, C_c, of 1.1.

The maximum and minimum densities of Oklahoma sand were determined in accordance with ASTM standards D4253 and D4254 respectively. The maximum density was 113 pcf and the minimum density was 97 pcf, which corresponds to void ratios of 0.46 and 0.7 respectively.

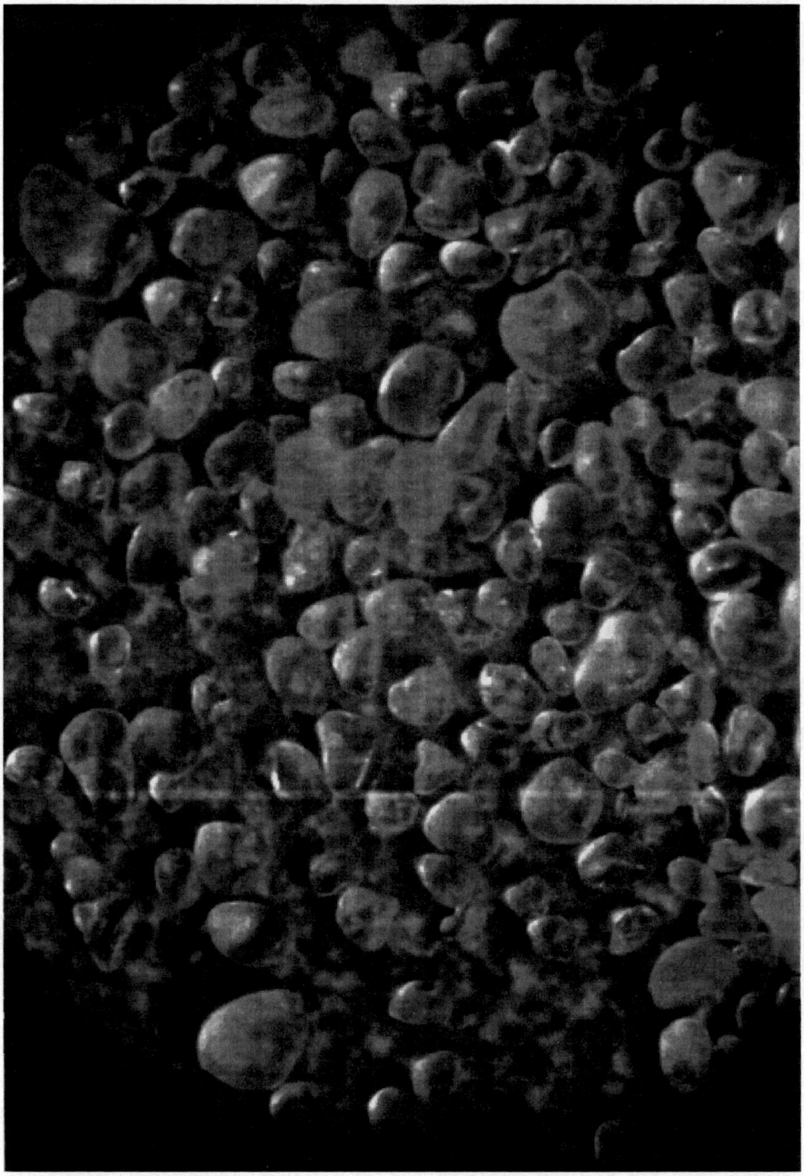

Fig. 7.2 Photomicrograph of Oklahoma Sand

7.4.2 Shear Strength

7.4.2.1 Triaxial Tests

Consolidated-drained triaxial tests were performed per Bishop and Henkel (1969). The specimens were 3 in. long and 1.5 in. in diameter. Dense specimens were compacted to an average density of 110 pcf, which is approximately the same placement density of the sand pluviated in the pressure chamber. Loose specimens had an average density of 100 pcf. The specimens were tested dry and were sheared at a deformation rate of 1.5 in./hour. The dense specimens reached their peak strength at a strain of 5% (Fig 7.3). No strain softening was observed for loose specimens (Fig. 7.4). Friction angles of 42° and 35° for dense and loose specimens, respectively, were obtained from the failure envelopes shown in Fig 7.5 and 7.6.

The strength of three saturated specimens matched that of dry specimens.

7.4.2.2 Direct Shear Tests

Direct shear tests were also performed in accordance with ASTM standard D3080-90 (Table 7.2). The samples were sheared at 0.08–0.1 inches/min. The angle of internal friction was stress dependent. Friction angles ranged between 37°–41° for low stress levels and 26°–31° for higher stress levels. Variations of density and saturation resulted in minor changes in the friction angle.

Malhotra (1991) obtained the angle of interfacial friction between sand and pile's material, δ, by filling the lower half of the direct shear box with a machined stainless steel (ASTM Type 304) plug. A δ of 26° was obtained for specimens having a relative density of 22%.

Triaxial tests are more reliable than direct shear tests, and should therefore be given more weight in evaluating the shear strength of Oklahoma sand.

7.4.3 Hydraulic Conductivity

The hydraulic conductivity was determined for a number of specimens compacted at a density of 110 pcf. Both flexible wall and fixed wall permeameters were used. The measured hydraulic conductivity averaged $1*10^{-3}$–$1*10^{-4}$ cm/sec.

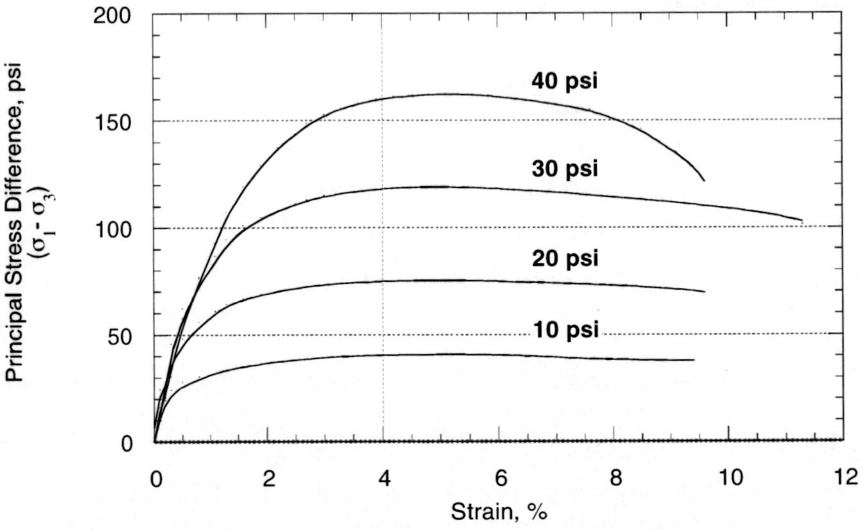

Fig. 7.3 Stress Strain Behavior of Dense, Dry, Oklahoma Sand

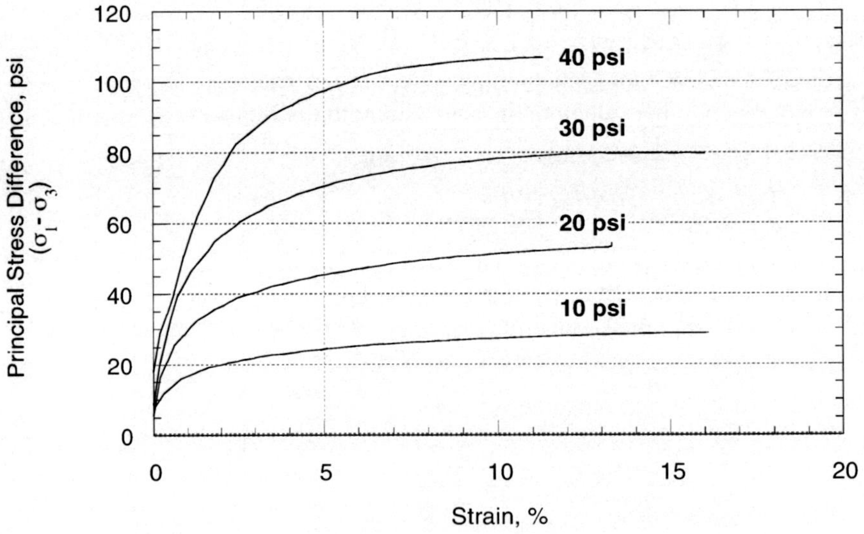

Fig. 7.4 Stress Strain Behavior of Loose, Dry, Oklahoma Sand

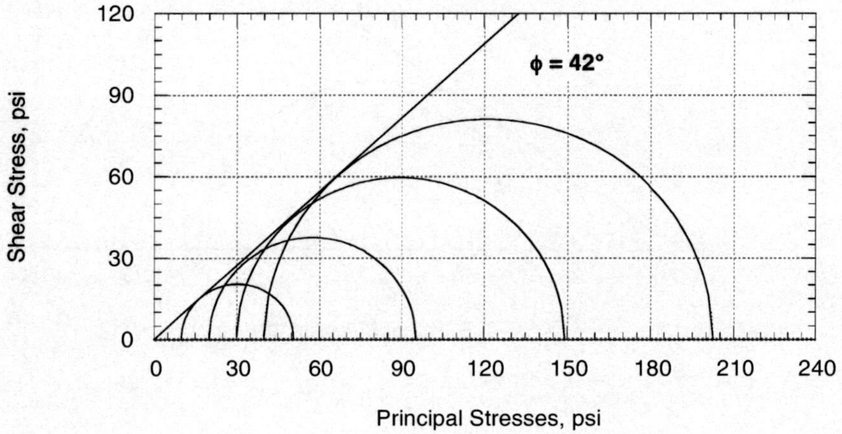

Fig. 7.5a Mohr-Coulomb Failure Envelope for Dense, Dry, Oklahoma Sand

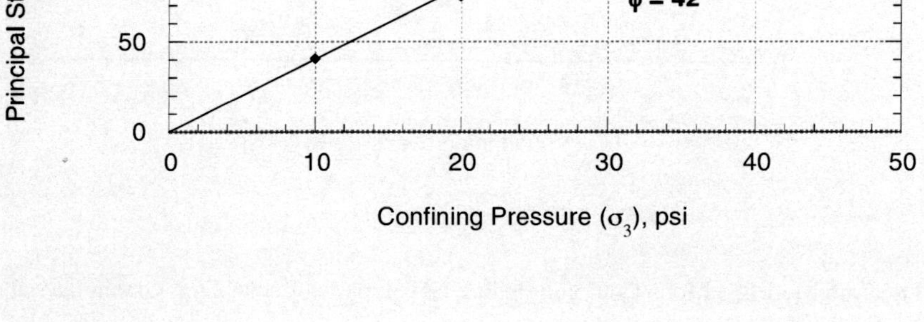

Fig. 7.5b Modified Mohr-Coulomb Failure Envelope for Dense, Dry, Oklahoma Sand

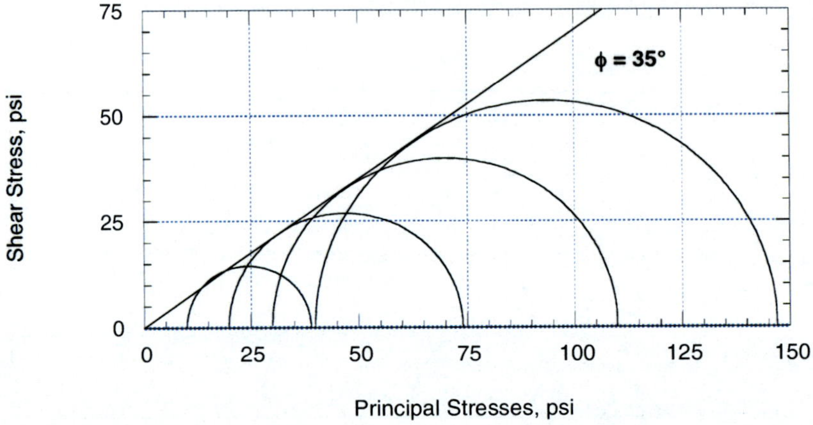

Fig. 7.6a Mohr-Coulomb Failure Envelope for Loose, Dry, Oklahoma Sand

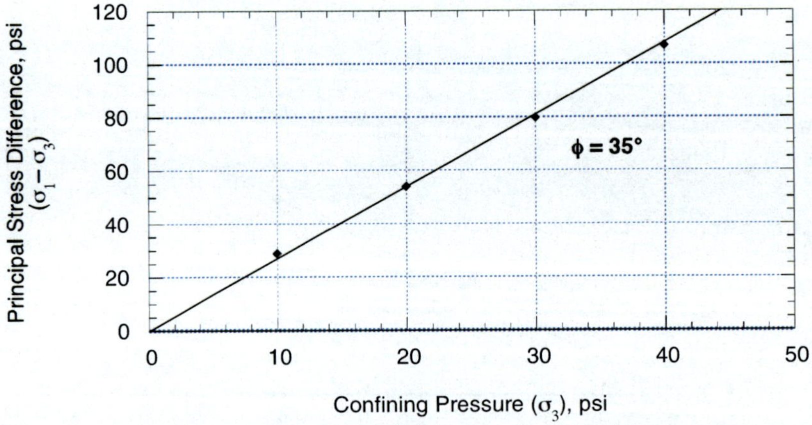

Fig. 7.6b Modified Mohr-Coulomb Failure Envelope for Loose, Dry, Oklahoma Sand

Table 7.2 Results of Direct Shear Tests on Oklahoma Sand

Source	Normal Stress	Dry/Wet	Density	ϕ
Malhotra 1990	2–6.5	Dry	D_r= 93%	41°
Malhotra 1990	2–6.5	Dry	D_r= 35%	38°
Pavlicek 1993	5–48.4	Dry	e= 0.53	31°
Jones 1994	0.15–4	Dry	Dense	37°
Ogilvie 1995	5–30	Dry	Loose	26°
Ogilvie 1995	5–30	Saturated	Loose	28°
Ogilvie 1995	5–30	Dry	Dense	30°
Ogilvie 1995	5–30	Saturated	Dense	28°

7.4.4 Changes in Sand Properties Due to Recycling

The sand used in these experiments was recycled between tests. A number of experiments were performed to determine if the recycling process changed the properties of the test sand. Grain size distribution as well as maximum and minimum density tests were performed on samples taken from a batch of sand that was recycled and used for approximately 15 times in load tests. No measurable change in grain size distribution was detected (Fig 7.7). The minimum and maximum densities of the recycled sand were within 0.5% of new sand.

Grain size distribution tests were also performed on samples recovered from the soil core (plug) inside the pile and from samples recovered from the soil-pile interface outside the pile. The soil core was retained inside the pile in all load tests that were performed on saturated sand, and the core was readily available. The sand at the external soil-pile interface was recovered after pile installation by pushing a thin walled 4.5-in.-diameter tube around the 3.5-in.-diameter pile. Sand was excavated from outside the recovery tube, and a plastic plate was pushed under the recovery tube and pile to cap them. Next the recovery tube was turned side ways with the pile inside it. The sand recovered from the soil-pile interface appears to be slightly coarser than new or recycled sand (Fig 7.7). This experimental anomaly may be explained as follows. Fine metal "dust" is formed due to the frictional resistance between the pile and soil. This dust is readily detectable by a discoloration of the sand at the soil-pile interface. This metal dust is mixed with the sand that was recovered from the soil-pile interface and results in increasing the apparent weight of the sand retained on some sieves. It is therefore believed that there is no difference in the grain size distribution of new sand and that recovered from the soil-pile interface.

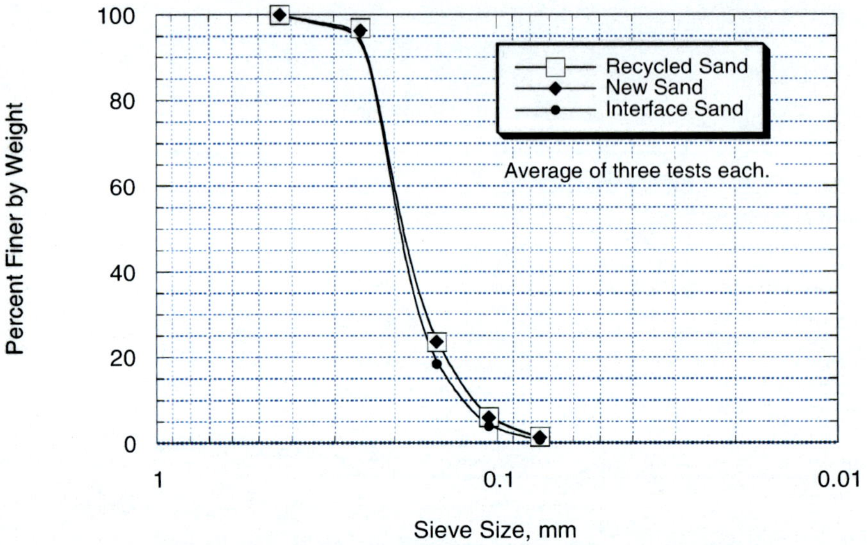

Fig. 7.7 Grain Size Distribution of New, Recycled, and Interface Sand

References

ASTM, Annual Book of Standards, Vol. 4.08, ASTM 1916 Race St., Philadelphia, PA 19103 (1993)

Bishop, A.W., Henkel, D.J.: The Measurements of Soil Properties in the Triaxial Test, 2nd edn. Edward Arnold, London (1969)

Bolton, M.D., Lau, C.K.: Scale Effects Arising from Particle Size. In: Proc. Centrifuge 1988, Balkema, Rotterdam (1988)

Holubec, I., D'Appolonia, E.: Effect of Particle Shape on the Engineering Properties of Granular Soils, Evaluation of Relative Density and its Role in Geotechnical Projects Involving Cohesionless Soils. In: ASTM STP 523, American Society of Testing and Materials, pp. 304–318 (1973)

Ishibashi, I.: Effect of Grain Characteristics on Liquefaction Potential – In Search of Standard Sand for Cyclic Strength. Geotechnical Testing Journal, GTJODJ 8(3) (1985)

Jones, W.C.: Axial Capacity of Suction Piles in Sand, M.Sc. Thesis, The University of Texas at Austin (1994)

Malhotra, S.: Development of a Model Pile-Driving Facility, M.Sc. Thesis, The University of Texas at Austin (1991)

Ogilvie, A.: Measurement of Engineering Properties for Sand Used in Tank Simulation of Offshore Pile Driving, Project Report, NSF Careers for Minority Scholars, The University of Texas at Austin (1995)

Pavlicek, R.W.: Axial Tensile Load Capacity of Suction Piles, M.Sc. Thesis, The University of Texas at Austin (1993)

8. Similitude between Model and Full Scale Piles

Abstract. This chapter presents an introduction to similitude and the scaling laws involved in modeling pile driving. The capabilities and limitations of the developed apparatus in modeling the installation and behavior of pile foundations are presented.

8.1 Introduction

Model experiments have proven to be useful in studying a wide range of geotechnical problems. The primary difficulty in experimental modeling of soil-structure interaction problems, such as pile driving, is the dependency of the soil strength on the stress level. This difficulty can be overcomed using a geotechnical centrifuge (*e.g.* Schofield 1980), a hydraulic gradient (*e.g.* Zelikson 1969), or using a pressure (calibration) chamber as is done in this research. In all three cases, careful consideration must be given to the scaling relationship between the model and the prototype in order to guarantee the integrity of experimental data.

A scaled model test is one in which the primary units of time, length, and mass are properly scaled, and from which scaled engineering data can be obtained directly. The proportionality between the model and the prototype is known as *Similitude* (Huntley 1952). Experiments on small prototypes do not necessarily satisfy scaling laws. They cannot be used to predict prototype behavior directly, but nevertheless can be used to verify analytical models.

This chapter presents the capabilities and limitations of the developed apparatus in modeling the installation and behavior of pile foundations.

8.2 Similitude

A prototype and its model are said to be in similitude if a point to point correspondence exists between the model and prototype (Langhar 1960). In geometrical terminology, corresponding points on a prototype and its model are referred to as *homologous* points. For transient problems, homologous times must also exist. The concept of similarity extends to many characteristics beside geometry and time. For example, it may be necessary that the mass distribution in a model be similar to that of a prototype.

M. Iskander: Behavior of Pipe Piles in Sand, SSGG, pp. 187–194.
springerlink.com

8.3 Scaling Laws

Scaling laws between models and prototypes may be obtained using either dimensional analysis or evaluating differential equations. Dimensional analysis is preferred to differential equations because it does not presuppose the relationship between the various variables involved (Cheney and Fragaszy 1984). Langhaar (1960 p.1) defined dimensional analysis as "*a method by which we deduce information about a phenomenon from the single premise that the phenomena can be described by a dimensionally correct equation among certain variables.*" As the number of variables affecting the problem increases, it becomes progressively more difficult to determine the function which relates the various variables involved. Buckingham π theorem is used to overcome this difficulty by grouping dependent and independent variables into a smaller number of non-dimensional products (Murff 1990). These non-dimensional products are referred to as π terms.

Gills (1988) used the variables shown in Table 8.1 to obtain the π terms shown in Table 8.2 for the variables involved in modeling of pile driving problems.

Scaling laws which satisfy similitude are derived by equating all the dimensionless π terms in both the prototype and the model as follows:

$$(\pi_i)\, model = (\pi_i)\, prototype \tag{8.1}$$

The scaling laws for piles driven in a centrifuge have been obtained by Gills (1988), Mehle (1989), and Allard (1990), among others, and are shown in Table 8.3.

8.4 Modeling Capabilities of the Experimental Facilities

The experimental apparatus developed in this study should be considered primarily as a small prototype that is used to study pile behavior. Nevertheless, the scaling relationships given in Tables 8.2–8.3 may be used to identify the modeling capabilities of the developed apparatus.

Table 8.1 Variables Used in Dimensional Analyses (Gill 1988)

Variables	Symbol	Dimensions
Independent		
Length	L	L
Density	ρ	M/L^3
Modulus	E	M/LT^2
Dependent		
Stress	σ	M/LT^2
Acceleration	a	L/T^2
Impact Energy	E_n	ML^2/T^2
Impulse	I	ML/T
Cohesion	C	M/LT^2
Dynamic Time	T	T
Wave Speed	P	L/T
Yield Strength	F	M/LT^2
Displacement	D	L
Area	A	L^2
Permeability	K	L/T
Hydrodynamic Time	T_h	T^2

Table 8.2 Non-Dimensional π Terms (Gill 1988)

Term	Property	Symbol	Expression	Dimensions
π_1	Stress	σ	$\frac{\sigma^2}{E^2}$	$\frac{(M/LT^2)^2}{(M/LT^2)^2}$
π_2	Acceleration	a	$\frac{a^2D^2\rho^2}{E^2}$	$\frac{(L/T^2)^2L^2(M/L^3)^2}{(M/LT^2)^2}$
π_3	Impact Energy	E_n	$\frac{E_n{}^2}{E^2D^6}$	$\frac{M^2L^4/T^4}{(M^2/L^2T^4)L^6}$
π_4	Impulse	I	$\frac{I^2}{\rho D^6 E}$	$\frac{M^2L^2/T2}{(M/L^3)L^6(M/LT^2)}$
π_5	Cohesion	C	$\frac{C^2}{E^2}$	$\frac{(M/LT^2)^2}{(M/LT^2)^2}$
π_6	Dynamic Time	T	$\frac{ET^2}{D^2\rho}$	$\frac{M/LT^2(T^2)}{(L^2)M/L^3}$
π_7	Wave Speed	P	$\frac{\rho P^2}{E}$	$\frac{M/L^3(L^2/T^2)}{M/LT^2}$
π_8	Yield Strength	F	$\frac{F^2}{E^2}$	$\frac{M^2/L^2T^4}{M^2/L^2T^4}$
π_9	Displacement	D	$\frac{D^2}{L^2}$	$\frac{L^2}{L^2}$
π_{10}	Area	A	$\frac{A^2}{D^4}$	$\frac{L^4}{L^4}$

Table 8.3 Scaling Relationships

Property	Prototype	Model
Length	n	1
Acceleration	1	n
Energy	n^3	1
Impulse	n^3	1
Force	n^2	1
Static Time	1	1
Dynamic Time	n	1
Dissipation Time	n^2	1
Density	1	1
Modulus	1	1
Stress	1	1
Strain	1	1
Velocity	1	1
Yield Strength	1	1

8.4.1 Modeling of Earth Stresses in the Pressure Chamber

In situ stresses are scaled in a pressure chamber by application of boundary stresses that simulate the effective stresses at the depth being modeled. For example, referring to Eq. 8.1 and the non-dimensional term π_1 (Table 8.2), a layer of saturated sand 100 ft below the surface, with anisotropic stress conditions (K_O = 0.5) may be represented by applying vertical and horizontal boundary stresses of 35 and 17.5 psi, respectively. The sand layer modeled in the pressure chamber is subject to two limitations. First, only one finite depth (ΔZ) could be studied at a time. The pressure chamber does not model the gradual increase in earth stress with depth, like the centrifuge. Second, although the effective stresses within the layer are modeled correctly, pore water pressures are not. The centrifuge suffers from the same discrepancy which is believed to be inconsequential for this problem.

The pressure chamber can apply pressures up to 40 psi in both the vertical and lateral directions. These stresses correspond to the stresses which occur in a soil layer with isotropic stress conditions (K_O = 1) 120 ft below the surface.

8.4.2 Modeling of Pile Driving Using the Hammer

In modeling the hammer operating speed, a conflict arises in time scaling. Pile driving is predominantly a dynamic phenomenon in which time is scaled by n, but the generation and dissipation of pore water pressures due to driving is a diffusive phenomenon in which time is scaled by n^2 (Table 8.3). The most practical solution to this problem is to model the hammer speed dynamically, and alter the rate of diffusion. The rate of diffusion can be slowed by using finer, less permeable, soils or a more viscous pore fluid. Each of these phenomena is discussed next.

8.4.2.1 Modeling of Dynamic Phenomena

Referring to Eq. 8.1 and the non-dimensional term π_6, similitude between the model pile hammer and production pile hammer requires that:

$$\pi_6 = \left(\frac{ET^2}{D^2\rho}\right)_m = \left(\frac{ET^2}{D^2\rho}\right)_p \tag{8.2}$$

where:

E	modulus of elasticity of pile material
ρ	density of pile material
D	pile diameter
T	dwell time of contact between the ram and anvil (rise time)
m,p	subscripts referring to model and prototype, respectively

The modulus of elasticity, E, and density, ρ, of the model and prototype piles are equal. Accordingly, Eq. 8.2 becomes:

$$\frac{T_m}{T_p} = \frac{D_m}{D_p} \tag{8.3}$$

As discussed earlier in Sections 6.5.2 and 6.8.22, the ram's dwell (rise) time is approximately 0.35 msec. Typical production hammers have dwell times of approximately 3–10 msec. Considering the 3.5 in.-diameter model piles tested in the chamber, the apparatus satisfies the requirement of homologous dynamic time (time similitude) for 30–100 in.-diameter piles. The dwell time of the ram could easily be increased, using thicker pile cushions, in order to satisfy similitude for smaller diameter piles. Commercial pile hammers deliver their energy to the pile in a duration that is usually shorter than the time taken by the energy wave to travel down through the pile and reflect back to the top (*2L/c*). A longer dwell time, however, would violate the requirements of impulse driving.

8.4.2.2 Modeling of Diffusive Phenomena

Terzaghi's (1943) time factor is the non-dimensional term applicable in diffusive problems (O'Neill 1991 and Poorooshasb 1990). Referring to Eq. 8.1, similitude between a model and a prototype would require that:

$$T = \left(\frac{c_r t}{R^2}\right)_m = \left(\frac{c_r t}{R^2}\right)_p \tag{8.4}$$

$$\frac{t_m}{t_p} = \frac{(c_r)_p}{(c_r)_m}\left(\frac{D_m}{D_p}\right)^2 \tag{8.5}$$

where:

c_r	radial coefficient of consolidation
D	pile diameter
R	pile radius
t	dissipation time
m,p	subscripts referring to model and prototype respectively

The coefficient of radial consolidation is given by Terzaghi (1943) as:

$$c_r = \frac{K}{m_v \gamma_w} \tag{8.6}$$

where:

K	coefficient of hydraulic conductivity
m_v	coefficient of volume compressibility
γ_w	unit weight of water

For a model and a prototype of equal compressibility (m_v) the coefficient of consolidation (c_r) is substituted in Eq. 8.5 such that:

$$\frac{t_m}{t_p} = \frac{K_p}{K_m}\left(\frac{D_m}{D_p}\right)^2 \tag{8.7}$$

The sand used in this study has a hydraulic conductivity which is one to two orders of magnitude smaller than that of offshore sands. The operating speed of the model pile hammer is 72 bpm which is equivalent to that of production pile hammers. Considering the 3.5 in.-diameter model piles used in this study, the developed apparatus satisfies the requirement of homologous diffusion time for piles 11–35 in. in diameter.

8.5 Conclusions

The experimental apparatus developed in this study should be considered primarily as a small prototype which is used to study pile behavior. Scaling relationships were used, however, to identify the modeling capabilities of the developed apparatus.

The developed apparatus satisfies most, but not all, the requirements of similitude for model tests. The primary inadequacy of the apparatus is its inability to model the gradual increase in soil stresses with depth, which limits the modeling capabilities to modeling one layer at a time. The pressure chamber can model the effective stresses in soil strata up to 120 ft below the mud line. The hammer can model both dynamic and diffusion times of 35 in.-diameter piles.

References

Allard, M.A.: Soil Stress Field Around Driven Piles, Ph.D. Dissertation, California Institute of Technology (1990)

Cheney, J.A., Fragaszy, R.J.: The Centrifuge as a Research Tool. Geotechnical Testing Journal, GTJODJ 7(4), 182–187 (1984)

Gills, J.J.: Development and Testing of a Device Capable of Placing Model Piles by Driving and Pushing in the Centrifuge, Ph.D. Dissertation, Univ. of Florida (1988)

Huntley, H.E.: Dimensional Analysis. MacDonald & Co (Publishers) LTD., London (1952)

Langhar, H.L.: Dimensional Analysis and Theory of Models, 4th edn. John Wiley & Sons, NY (1960)

Mehle, J.S.: Centrifuge Modeling of Pile Driving, Master Thesis, Department of Civil, Environmental, and Architectural Engineering, University of Colorado, Boulder Colorado (1989)

Murff, J.D.: Marine Foundation Engineering Class Notes, Texas A&M University (1990)

O'Neill, M.: Houston Calibration Chamber: Case Histories. In: Proc. 1st Int. Symposium on Calibration Chamber Testing, Potsdam, NY, pp. 277–288. Elsevier, NY (1991)

Poorooshasb, F.: On Centrifuge Use for Ocean Research. Marine Geotechnology 9, 141–158 (1990)

Schofield, A.N.: Cambridge Geotechnical Centrifuge, 20th Rankine Lecture. Geotechnique 30(3), 227–268 (1980)

Terzaghi, K.: Theoretical Soil Mechanics. Wiley, NY (1943)

Zelikson, A.: Geotechnical Models Using The Hydraulic Gradient Similarity Method. Geotechnique 19(4), 495–508 (1969)

9. Load Tests Using the Double–Wall Pipe Pile in Sand

Abstract. Typical data obtained using the developed double wall pile, pile hammer, pressure chamber, and loading apparatus is presented in this chapter. The data illustrates a number of important phenomena including pile plugging, development of pore water pressure, and load transfer during jacking (pushing), driving, and load testing.

9.1 Introduction

Approximately 50 load tests were performed using the developed facilities (Iskander 1995, Alansari 1999, and Malhotra 2007). The majority of these tests were performed using uninstrumented piles, mainly to verify the performance of the equipment, but also to study pile plugging. Tests were performed in both dry and saturated Oklahoma sand. Piles were installed using two methods; driving and pushing.

The load tests performed using the instrumented double-walled pile, discussed in chapter 5, provided far more data than the tests on uninstrumented piles. The data consisted of measurements of axial and radial stresses, as well as pore pressures, outside and inside the pile. Typical data obtained using the instrumented pile is presented in this chapter. The conclusions are, however, based on the author's experience gained from testing both instrumented and uninstrumented piles.

9.2 Test Procedure

The tests presented in this chapter were all performed on samples of freshly pluviated sand. For tests in saturated sand, 75% of the confining pressure was applied after pluviation. Next, deaired water was introduced at the base of the chamber and allowed to rise slowly, with a gradient not exceeding 0.5. Introduction of water presumably resulted in reorientation of the sand particles and the chamber required at least 4 hours to reach equilibrium. Next, the confining pressure was

M. Iskander: Behavior of Pipe Piles in Sand, SSGG, pp. 195–243.
springerlink.com

increased to its final value, which initiated a consolidation process[1]. Consolidation was complete over night and testing was performed the following day. For tests in dry sand, all the confining pressure was applied immediately after pluviation, and testing was performed the following day.

All the tests reported herein were performed in soil samples with isotropic stresses. The vertical and lateral confining pressures were set to 20 psi for the tests presented in this chapter. These stresses correspond to the stresses in layers of soil 25 and 50 ft below the surface for dry and saturated sand, respectively.

Piles were installed to a penetration depth of approximately 19 in. below the surface by either driving or pushing. Driving was performed at a rate of 60 bpm using the hammer discussed in chapter 6. Pushing was performed at a rate of 2500 lb/min using the load-controlled MTS feed-back control system discussed in chapter 4. The MTS was set to unload at a rate of 800 lb/min.

Compression load tests were performed immediately after installation by pushing. For piles installed by driving, compression load tests were performed approximately one hour after driving, which allowed any developed excess pore pressures to dissipate. Piles were loaded using the MTS system at a rate of 2500 lb/min, and for a distance of approximately 5 in.

Tensile pull-out tests were performed immediately after compression load tests. Tensile loads were applied at a rate of 2500 lb/min, using the MTS system.

Data was recorded at a rate of 0.5 Hz during installation by pushing, and during both compression load tests and tension pull-out tests. Each recorded data point represented the average of 40 measurements. During driving, data was recorded at a rate of 10 Hz, and no averaging was performed.

9.3 Data Presentation

9.3.1 Stresses and Pore Pressures

The instrumented double-wall pile used in this study consists of two concentric cylinders which are rigidly fixed at the top and are free to strain independent of one another elsewhere (Chapter 5). The pile is used to delineate the stresses acting outside the pile from those acting inside the pile, and measure each separately. Axial and circumferential strains as well as pore water pressures are measured at the locations indicated in Table 9.1. The measurements define the complete state

[1] Application of the full confining pressure, in one stage, after the sample is saturated results in liquefaction of the sand.

of stress acting against the pile. The pile is 3.5 in. in diameter, 33.25 in. long, and has a wall thickness of 0.315 in. The pile is designed to have a changeable tip shape. All the tests presented in this chapter were performed with a flat tip.

Pile installation and loading may results in instability of strain gage readings which is referred to as *zero-shift*. Zero-readings were measured before installation and zero-shifts were measured after pull-out. The data recorded during installation and compression load tests was reduced using the zero-readings measured before installation. The data recorded during pull-out tests was reduced using the zero-shifts measured after pull-out.

In this chapter, the measured total stresses and excess pore water pressures, which occur during installation and loading are presented. Axial load and skin friction distributions are also computed and presented.

Table 9.1 Instrumentation of the Double Wall Pile

Level	Height Above Tip, in.	Transducers
1	3	Strain and Pore Pressure
2	5	Strain and Pore Pressure
3	10	Strain and Pore Pressure
4	17	Strain and Pore Pressure
5	24	Strain

9.3.2 Plug Measurements

The elevation of the soil core (plug) inside the pile was measured throughout testing. Hvorslev (1949) and Paikowsky (1989), among others, presented measurements of plugging using the specific (incremental) recovery ratio (SRR) and the plug length ratio (PLR). These ratios are defined as follows:

$$SRR = \frac{Incremental\ Change\ in\ Plug\ Height}{Incremental\ Change\ in\ Penetration} \tag{9.1}$$

$$PLR = \frac{Incremental\ Plug\ Height}{Incremental\ Penetration} \tag{9.2}$$

Plugging is indicated when the specific recovery ratio approaches zero. A specific recovery ratio of 100% (or more) indicates that the pile is penetrating in a coring mode. Plugging is also indicated by a change in the slope of the PLR-penetration curve.

During pull-out, the soil core may slip or be retained. The specific slip ratio (SSR) and the plug slip ratio (PSR) are defined by the author as the reverse of the SRR and PLR ratios, as follows:

$$SSR = \frac{\textit{Incremental Change in Plug Slip}}{\textit{Incremental Change in Pull-Out Distance}} \tag{9.3}$$

$$PSR = \frac{\textit{Incremental Plug Slip}}{\textit{Incremental Pull-Out Distance}} \tag{9.4}$$

Retention of the plug during pull out is indicated by an SSR approaching zero or the slope of the PSR-penetration curve becoming horizontal. Complete slippage of the soil core is indicated by an SSR approaching 100%, or a PSR-penetration curve having a 45° slope.

9.4 Load Tests on the Pile Pushed (Jacked) in Dry Sand

Typical results during the installation and load test stage of the instrumented pile pushed in dry sand are presented in Fig. 9.1–9.9. The results of the pull-out tests are presented in Fig. 9.10–9.13.

9.4.1 Jacking (Pushing) Stage

During the first four diameters of pushing (14 inches), most of the capacity was due to end bearing, which was transferred to the pile top through the external pile wall (Fig. 9.1). As the pile penetrated deeper, skin friction developed on the internal and external surfaces of the pile (Fig. 9.1 and 9.2). Plugging occurred at a penetration of 18.5 in. and was readily identifiable by a sharp increase in the load transferred through the internal pile wall (Fig. 9.2), accompanied by a sharp decrease in the load transferred through the external pile wall (Fig. 9.1).

Plugging is also indicated by the specific recovery ratio (SRR) approaching zero at a penetration of 18.5 in. (Fig. 9.3). Another indicator of plugging is the change in the slope of the PLR-penetration curve at a penetration of 18.5 in. (Fig. 9.3).

The elevation of the soil *core* inside the pile was measured throughout the test and was used to confirm the formation of a plug (Fig. 9.3). As the pile penetrated into the ground, the elevation of the soil in the pile increased above the ground level outside the pile. A maximum elevation 5 in. above the ground level was reached at 14 in. penetration. The elevation of the soil core remained relatively constant as it developed skin friction against the pile. The core formed a *plug* at a penetration of 18.5 in. After plugging the surface of the plug moved downward at the same rate of pile penetration.

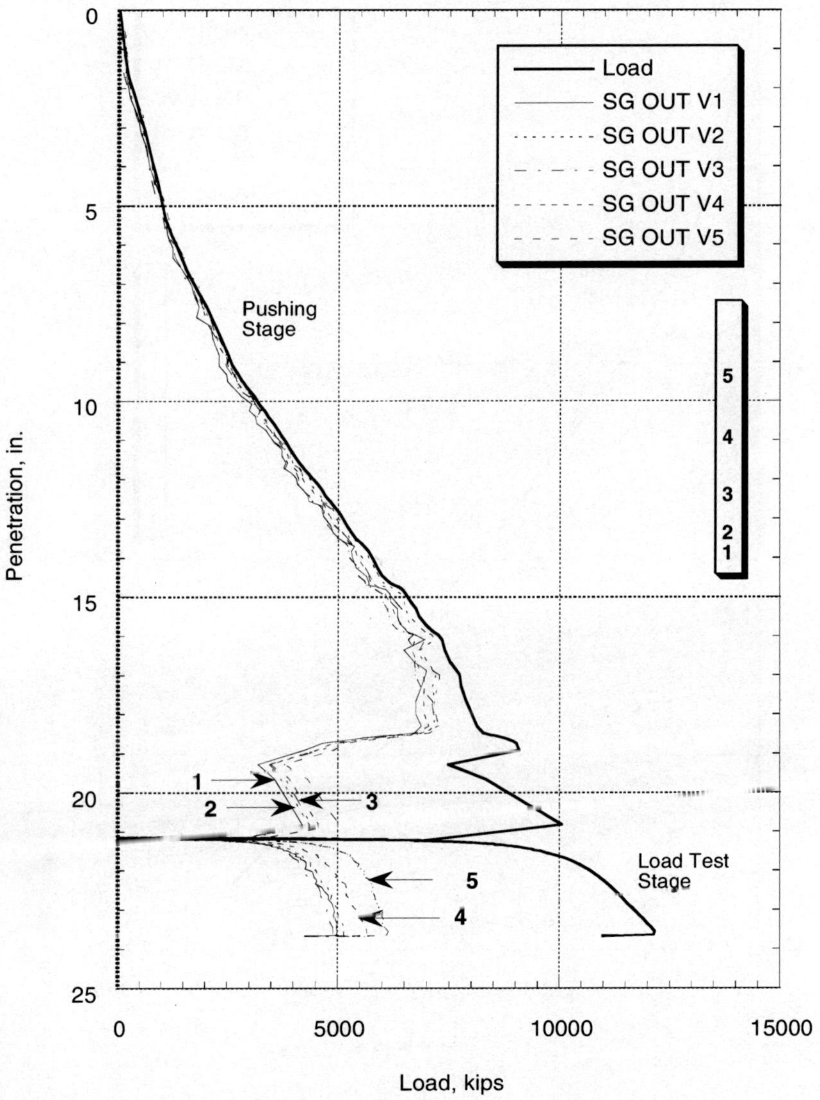

Fig. 9.1 Load-Settlement Curve of a Pipe Pile, Pushed in Dry Sand, Showing Axial Load Transfer Outside the Pile

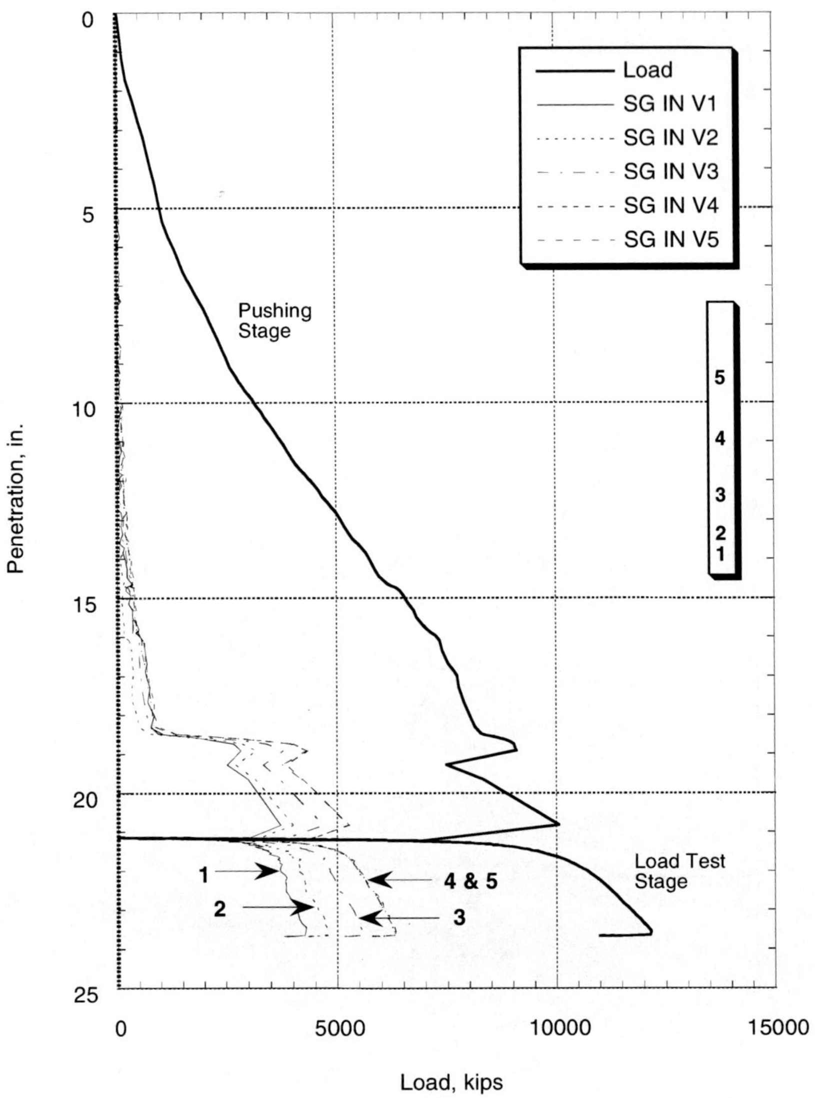

Fig. 9.2 Load-Settlement Curve of a Pipe Pile, Pushed in Dry Sand, Showing Axial Load Transfer Inside the Pile

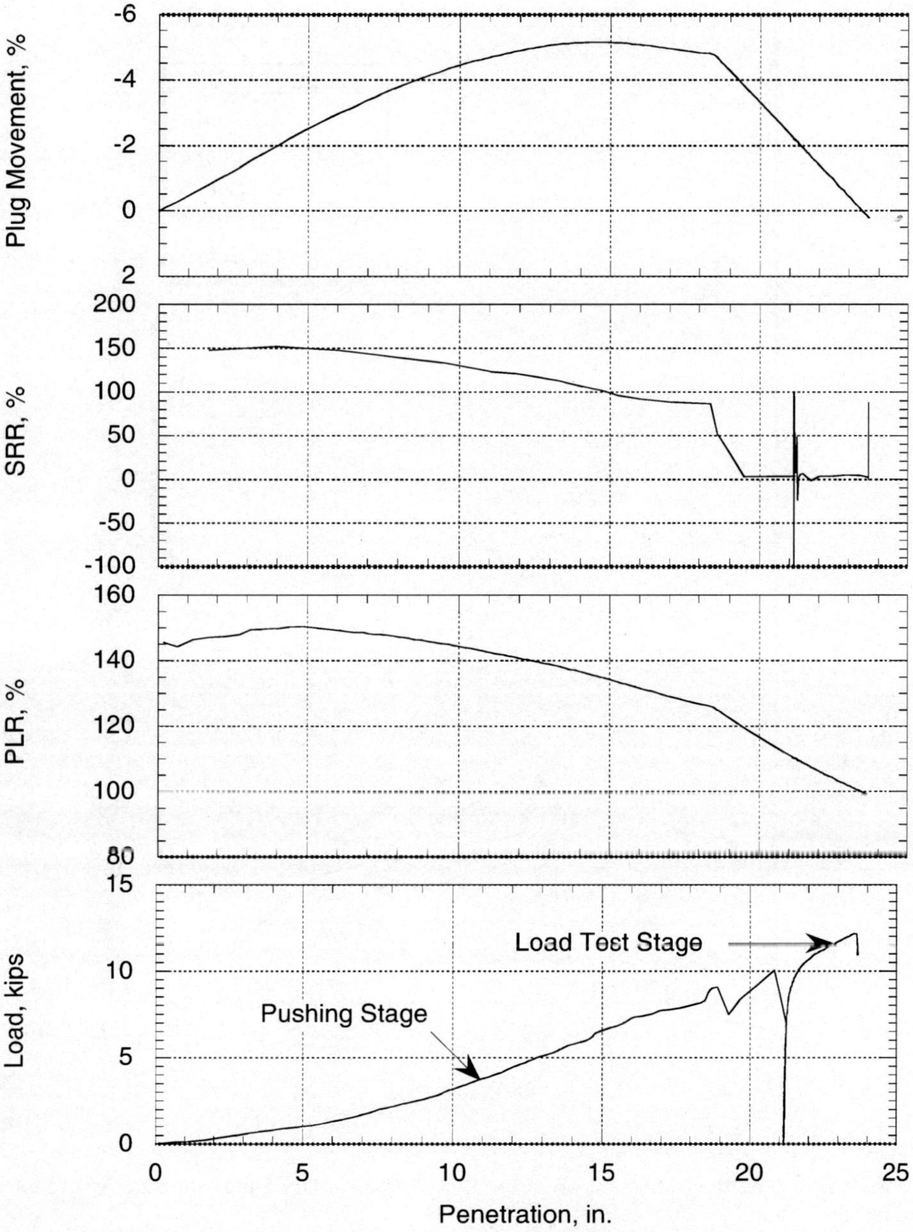

Fig. 9.3 Plug Movement, Specific Recovery Ratio (SRR), and Plug Length Ration (PLR) of a Pipe Pile Pushed in Dry Sand

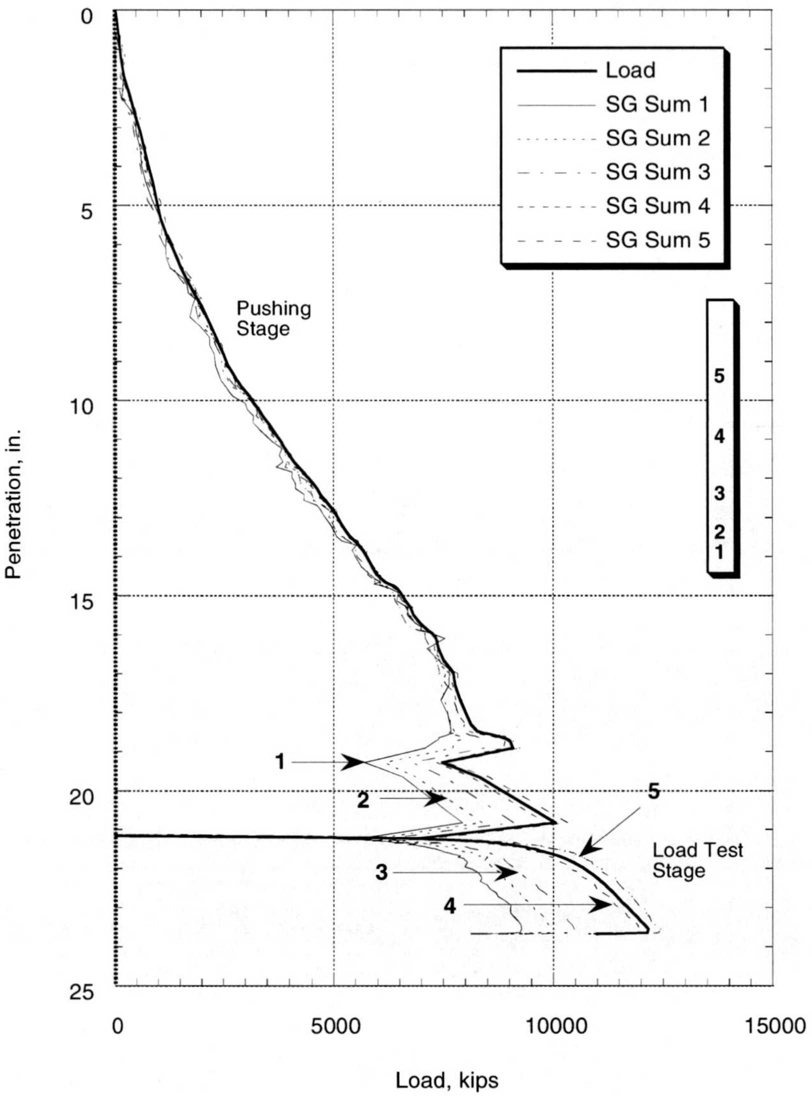

Fig. 9.4 Load-Settlement Curve of a Pipe Pile, Pushed in Dry Sand, Showing Axial Load Transfer

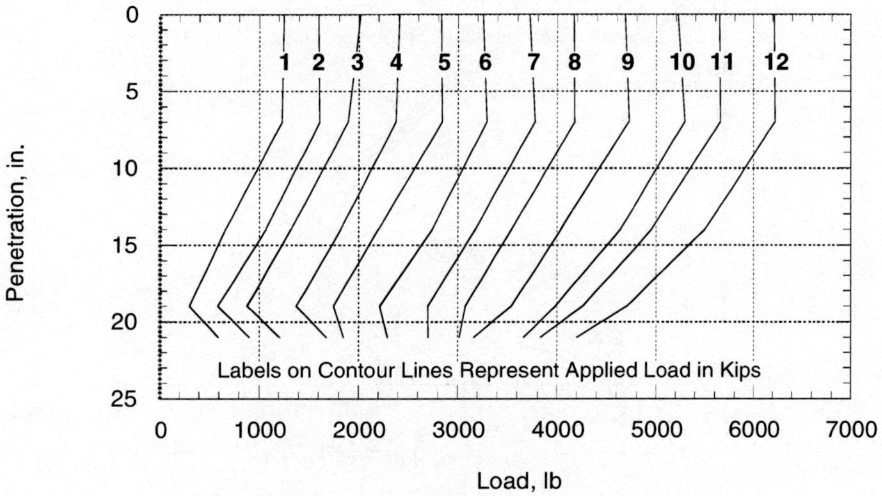

Fig. 9.5 Axial Load Distribution, during Load Test Stage, Outside a Pipe Pile Pushed In Dry Sand

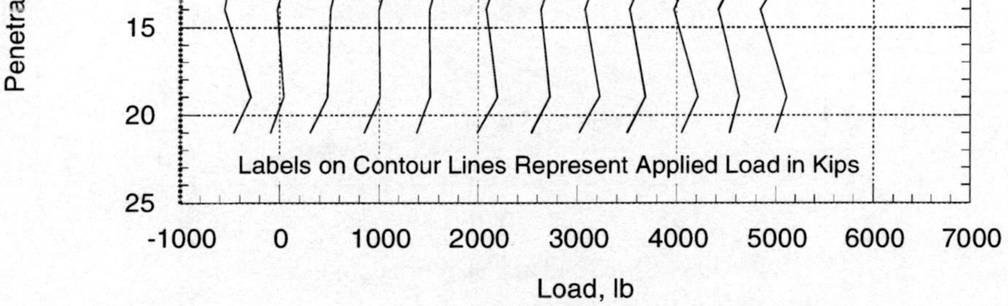

Fig. 9.6 Axial Load Distribution, during Load Test Stage, Inside a Pipe Pile Pushed In Dry Sand

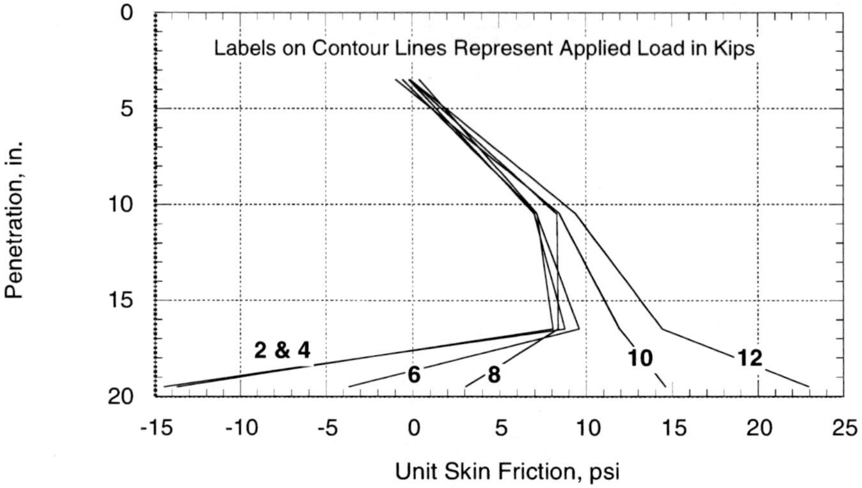

Fig. 9.7 Distribution of Unit Skin Friction, during Load Test Stage, Outside a Pile Pushed In Dry Sand

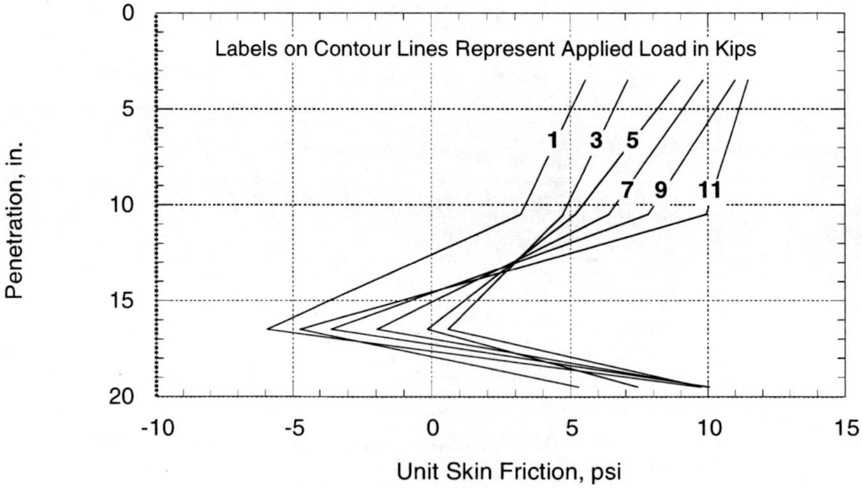

Fig. 9.8 Distribution of Unit Skin Friction, during Load Test Stage, Inside a Pile Pushed In Dry Sand

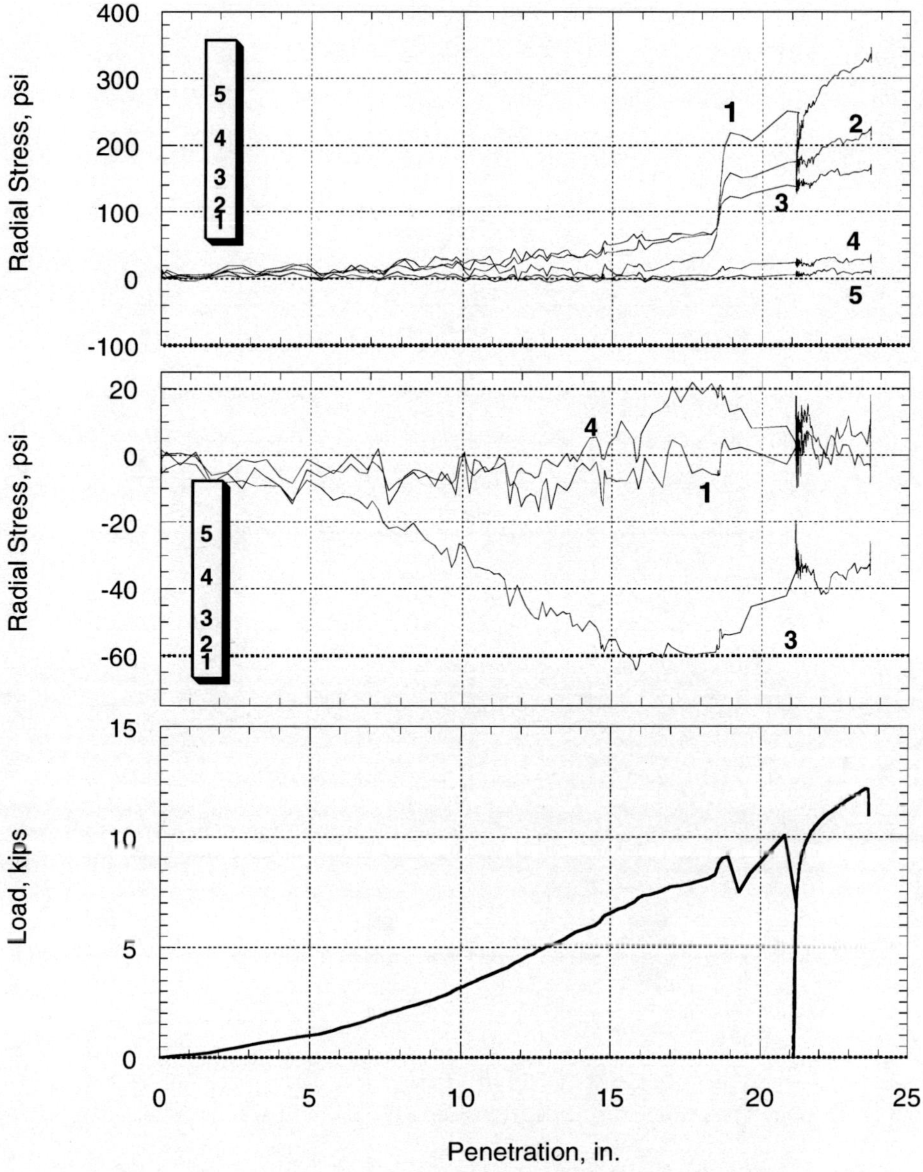

Fig. 9.9 Radial Stresses Acting on the Internal (Top) and External (Middle) Surfaces of a Pipe Pile Pushed (Bottom) in Dry Sand.

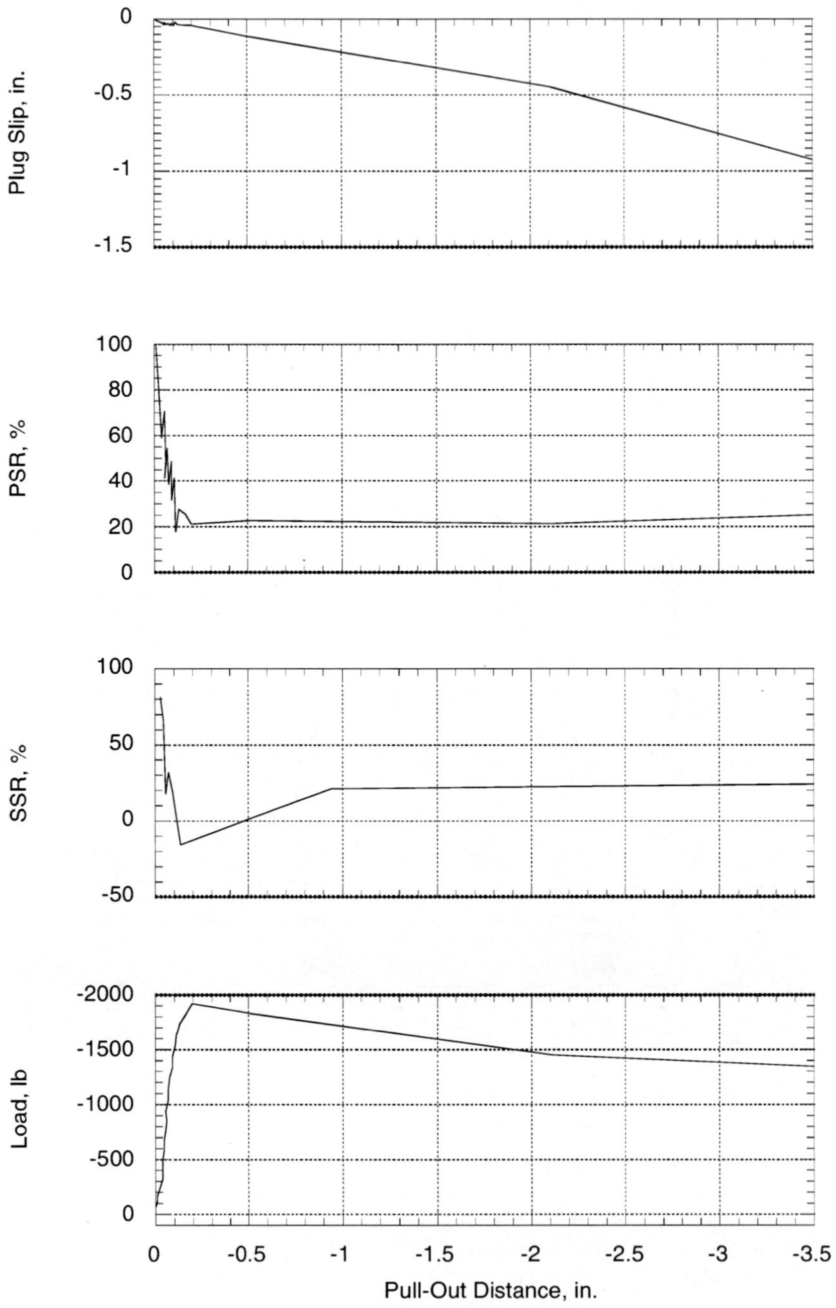

Fig. 9.10 Plug Slip, Plug Slip Ratio (PSR), and Specific Slip Ratio (SSR) during Pull-Out Test in Dry Sand

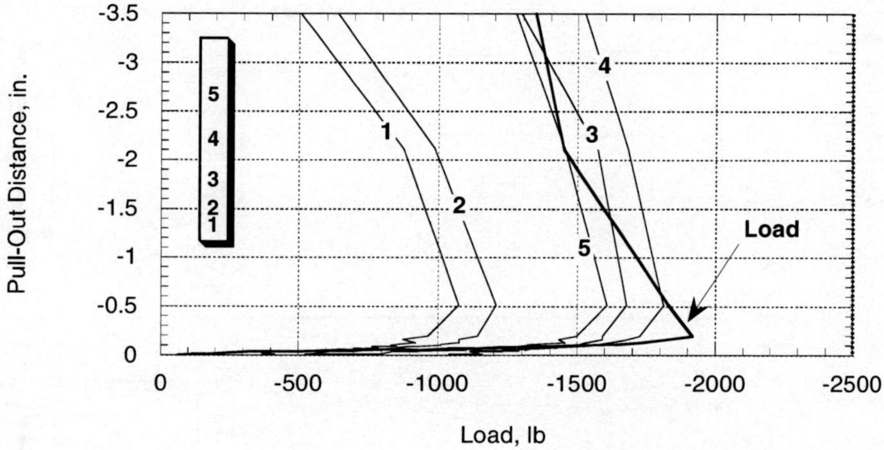

Fig. 9.11 Pull-Out Curve of a Pile, Pushed in Dry Sand, Showing Axial Load Transfer Outside the Pile

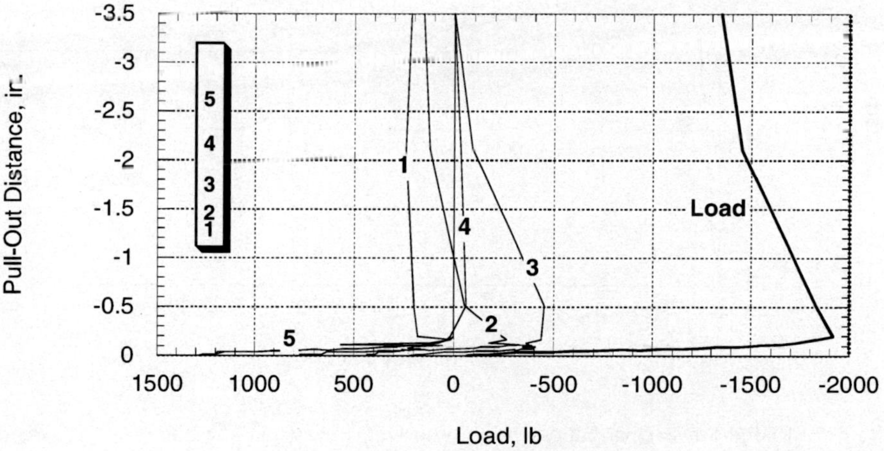

Fig. 9.12 Pull-Out Curve of a Pile, Pushed in Dry Sand, Showing Axial Load Transfer Inside the Pile

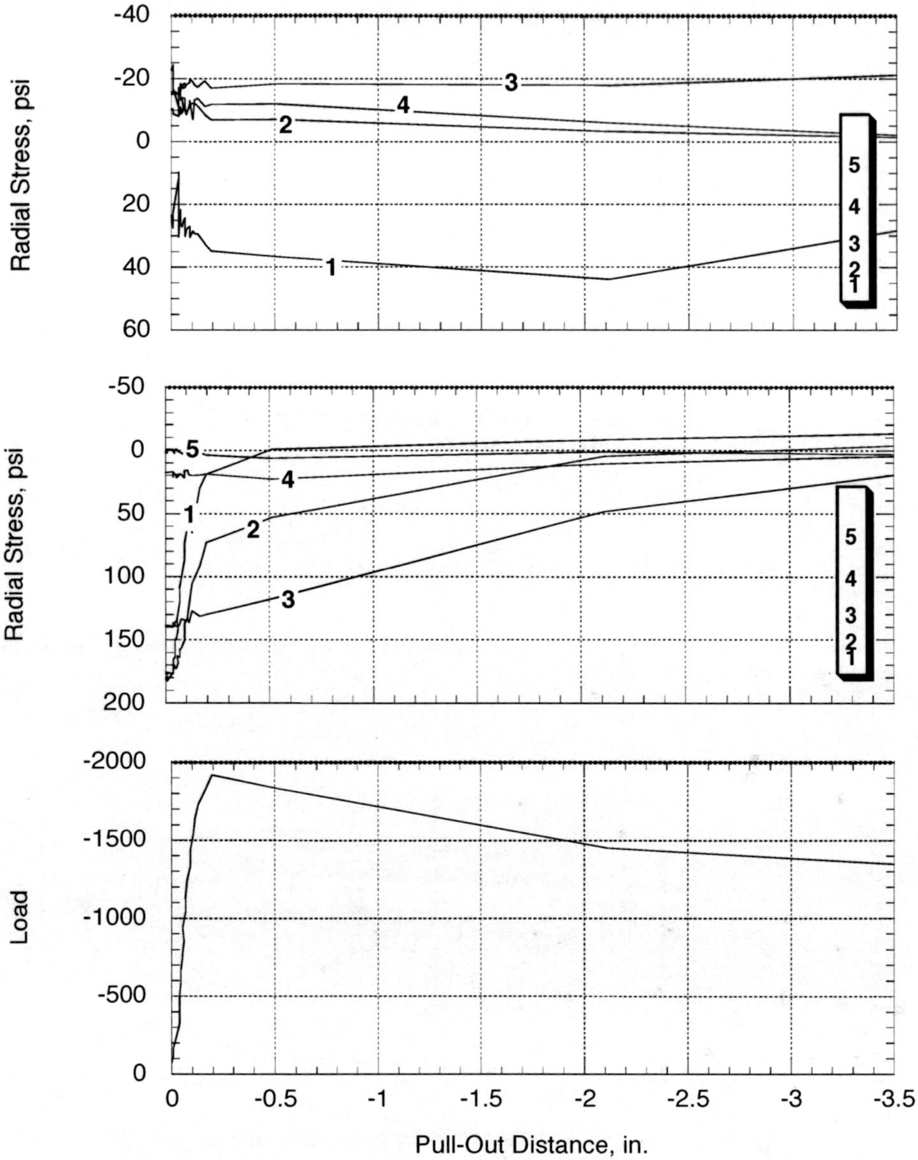

Fig. 9.13 Radial Stress Distribution During Pull-Out (a) Outside The Pile (Top), (b) Inside The Pile (Middle) , and (c) Load-Displacement Curve. (Bottom) for Pile Installed by Pushing in Dry Sand.

When plugging occurred, 80% of the end bearing was transferred to the internal pile wall in a length equal to one pile diameter (Fig. 9.2). The remaining 20% was transferred in the next diameter.

Figure 9.4 represents the load transfer obtained using a single wall pile. The figure was generated by summing corresponding axial loads measured on the internal and external pile walls. Comparison between Fig. 9.4 and Fig. 9.1–9.2 demonstrates the utility of the double-wall concept. Plugging can not be identified on Fig. 9.4. More importantly, the load transfer obtained using single-walled piles (Fig. 9.4) projects an apparent increase in skin friction and a decrease in end bearing over their real values. The good fit between the sum of the loads measured in the top levels of the external and internal pile walls (level 5) and the load measured at the pile top is indicative of the over all high quality of the collected data.

9.4.2 Load Test

The axial load distributions shown in Fig. 9.5–9.6 were used to compute the unit skin friction mobilized during the load test stage. Skin frictions were computed as the difference in the load measured at any two consecutive levels divided by the surface area of the pile between the two levels. The computed skin frictions were assumed to act at the mid height between the two levels used in the computation.

Outside the pile, initially large negative skin frictions occurred near the tip, probably due to residual stresses (Fig. 9.7). The negative skin friction values reduced as the applied load increased. Finally, the skin frictions near the tip became positive as the applied load used in the load test exceeded the maximum value used during pile pushing. Negative skin frictions have been observed by a number of researchers in the field, *e.g.* Al-Shafei (1994), and in the lab, *e.g.* Vesic (1967).

Inside the pile, large skin frictions developed near the tip. Large positive skin frictions were also measured near the top, accompanied by large negative skin frictions in the middle of the pile. The measured skin friction at the middle and top of the pile appear to be in error and may have resulted from a combination of a large stress gradient and the large distance between strain gages in the upper portion of the pile.

A better indication of the state of stress inside the pile is the radial stresses measured inside the pile (Fig. 9.9a). Radial stresses approached 300 psi near the pile tip and quickly reduced within three pile diameters. These stresses must have resulted from arching of the sand due to plugging.

Outside the pile, the measured radial stresses decreased from their initial values near the tip and top, but increased in the middle (Fig. 9.9b).

Before pile installation, the pressure chamber was set to a uniform isotropic stress of 20 psi. The dense sand used in this study has a friction angle of 42°. A uniform skin friction distribution of 18 psi is predicted outside the pile, based on the Mohr-Coulomb failure criteria. The measured skin friction was nevertheless triangular (instead of uniform) and the average skin friction was less than 10 psi. The triangular stress distribution may have resulted from the influence of the chamber's boundary conditions. Chamber effects, however, do not explain the reduced skin friction values. Pile installation may have resulted in a zone of dense sand encircling a zone of loose sand, around the pile, thus preventing by arching the development of full earth pressure (Robinsky and Morrison, 1964).

9.4.3 Pull-Out Test

Approximately 15% of the force required to push the pile was required to pull it out. The small tensile capacity is a result of the small length to diameter ratio of the test pile (10:1), which was required to accommodate the available head clearance[2]. The plug continued to slip during pull-out at approximately half of the pull-out rate (Fig. 9.10).

Pull-out capacity was derived primarily from friction outside the pile (Fig. 9.11–9.12). The quality of the measurements in tension are lower than those in compression due to the reduced resolution caused by the smaller loads. It is however clear that load transfer outside the pile followed the pull-out load (Fig. 9.11). It is also clear that the plug contained large residual stresses which were released within 0.1 in. of pull-out displacement (Fig. 9.12). The residual stresses inside the pile are compressive at the tip and top and tensile at the center, which tend to support the skin friction distribution measured during the load test (Fig. 9.8).

The measured radial stress distribution indicates that the arching stresses within the plug were released as the pile pulled out of the ground (Fig. 9.13). Small changes in radial stress occurred outside the pile during pull-out.

9.5 Load Tests on the Pile Pushed (Jacked) in Saturated Sand

The results of the load tests on piles pushed in saturated sand were in general agreement with piles pushed in dry sand. Typical results during the installation and load test stage of the instrumented pile pushed in saturated sand are presented in Fig. 9.14–9.23. The results of the pull-out test are presented in Fig. 9.24–9.28.

[2] Refer to chapters 4 and 5 for additional details.

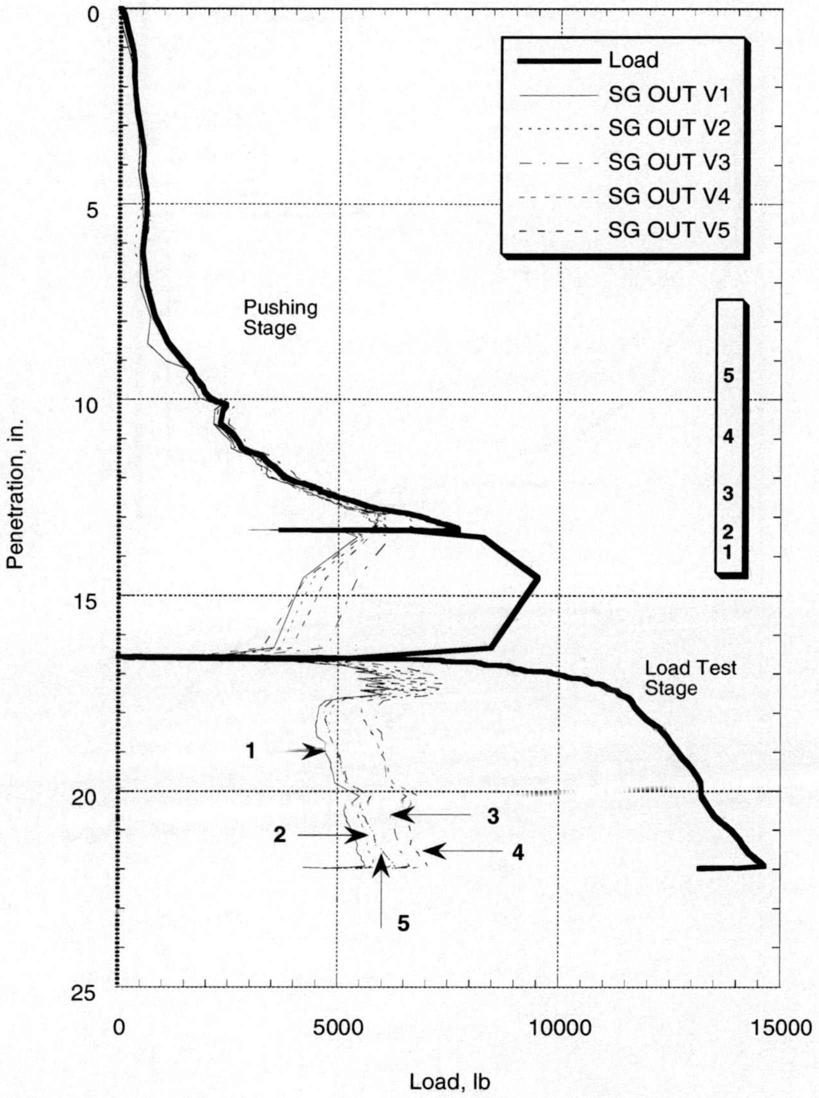

Fig. 9.14 Load-Settlement Curve of a Pile, Pushed in Saturated Sand, Showing Axial Load Transfer Outside the Pile

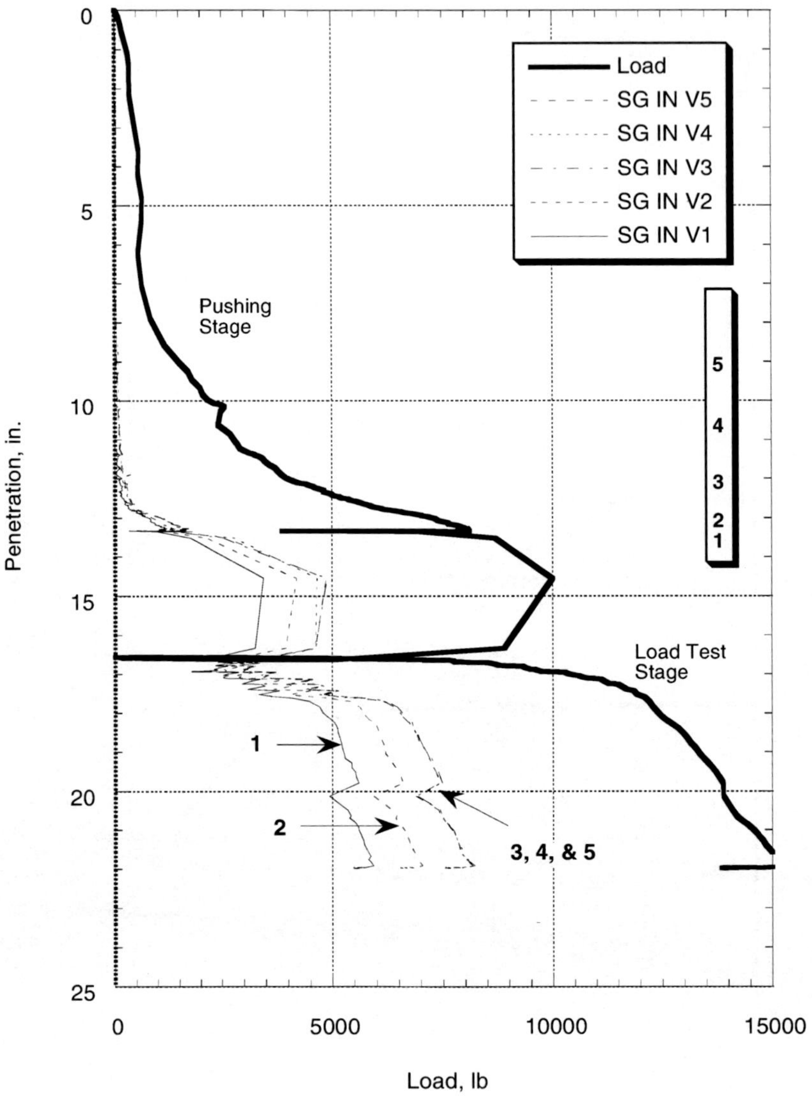

Fig. 9.15 Load-Settlement Curve of a Pile, Pushed in Saturated Sand Showing Axial Load Transfer Inside the Pile

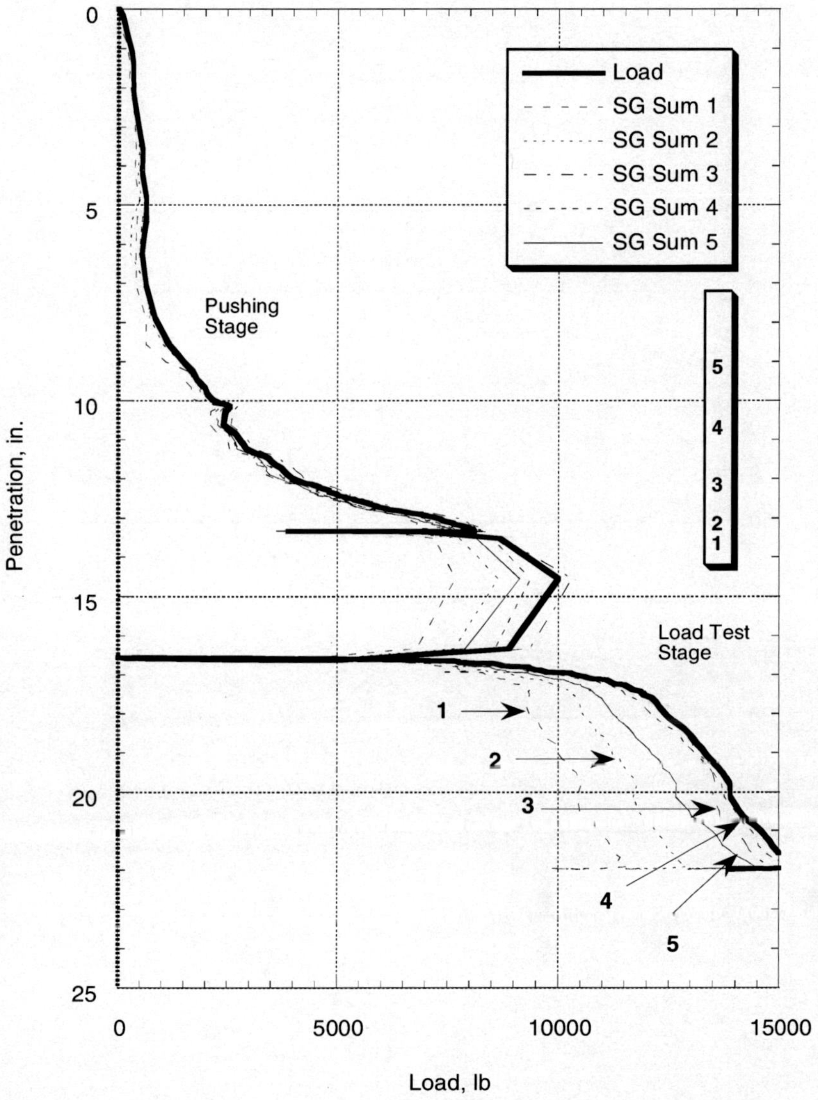

Fig. 9.16 Load-Settlement Curve of a Pile, Pushed in Saturated Sand, Showing Axial Load Transfer

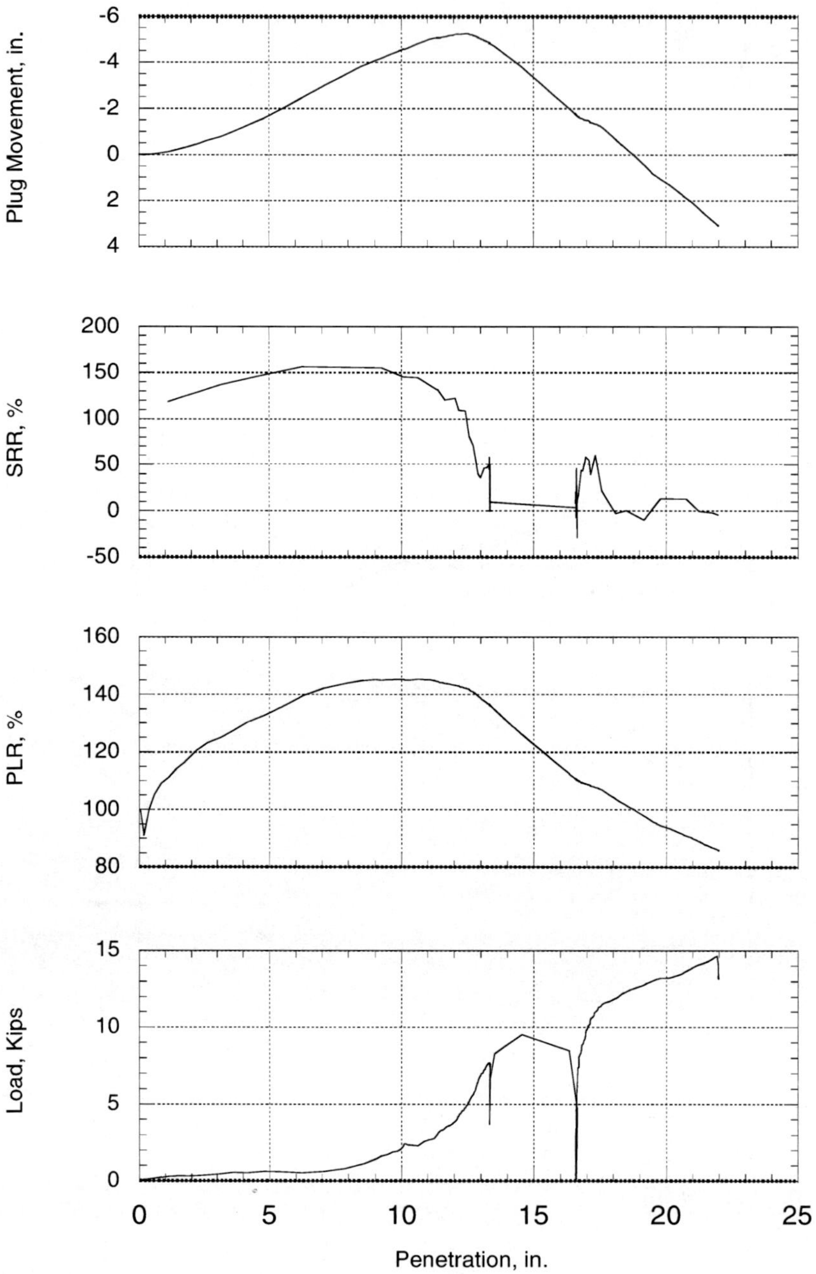

Fig. 9.17 Plug Movement, SRR, and PLR of a Pile Pushed in Saturated Sand

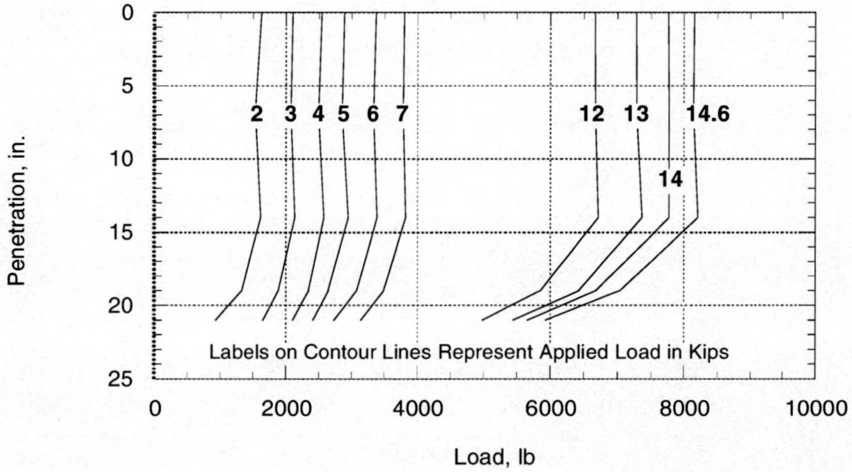

Fig. 9.18 Axial Load Distribution, During Load Test Stage, Outside a Pipe Pile Pushed in Saturated Sand

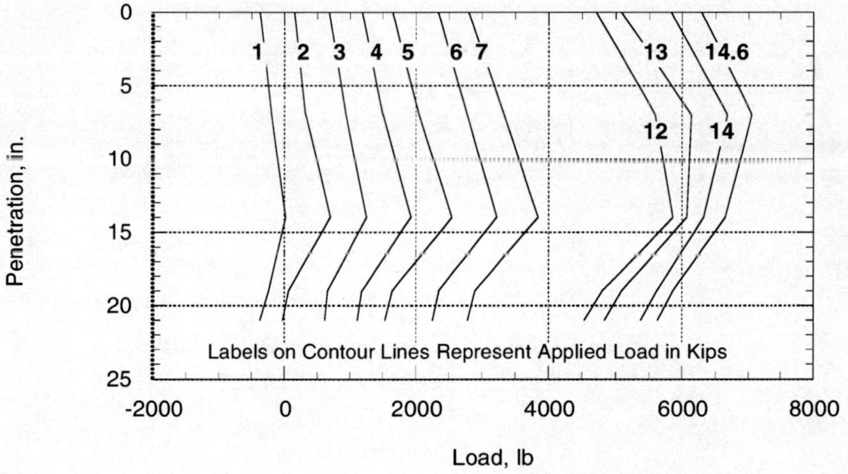

Fig. 9.19 Axial Load Distribution, During Load Test Stage, Inside a Pipe Pile Pushed in Saturated Sand

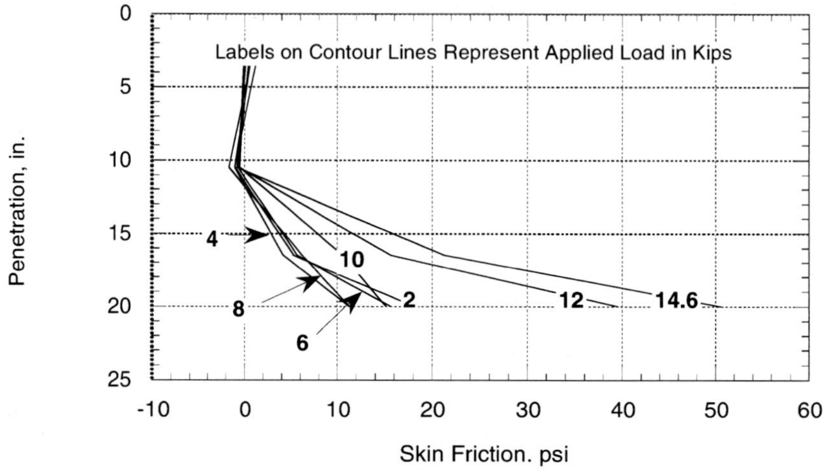

Fig. 9.20 Skin Friction Distribution, During Load Test Stage, Outside a Pipe Pile Pushed in Saturated Sand

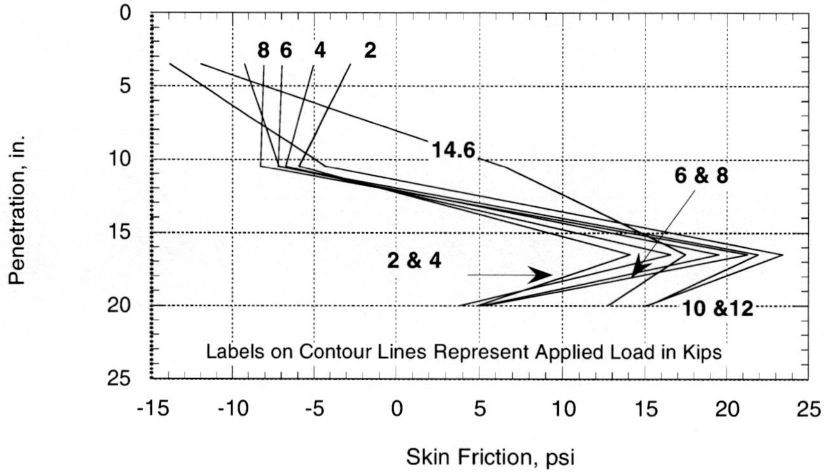

Fig. 9.21 Skin Friction Distribution, During Load Test Stage, Inside a Pipe Pile Pushed in Saturated Sand

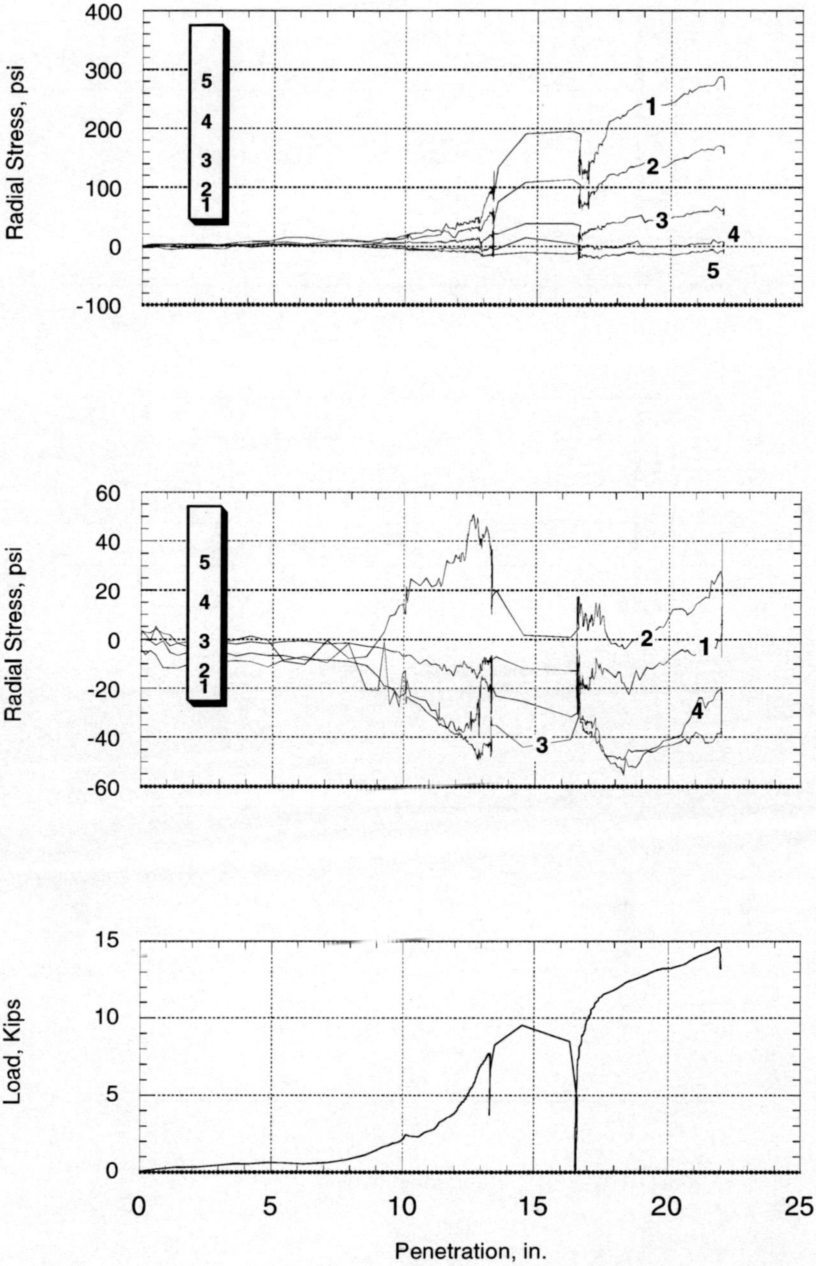

Fig. 9.22 Radial Stresses Acting on the (a) Internal (Top) and (b) External Surfaces (Middle) of a Pipe Pile (c) Pushed in Dry Sand (Bottom)

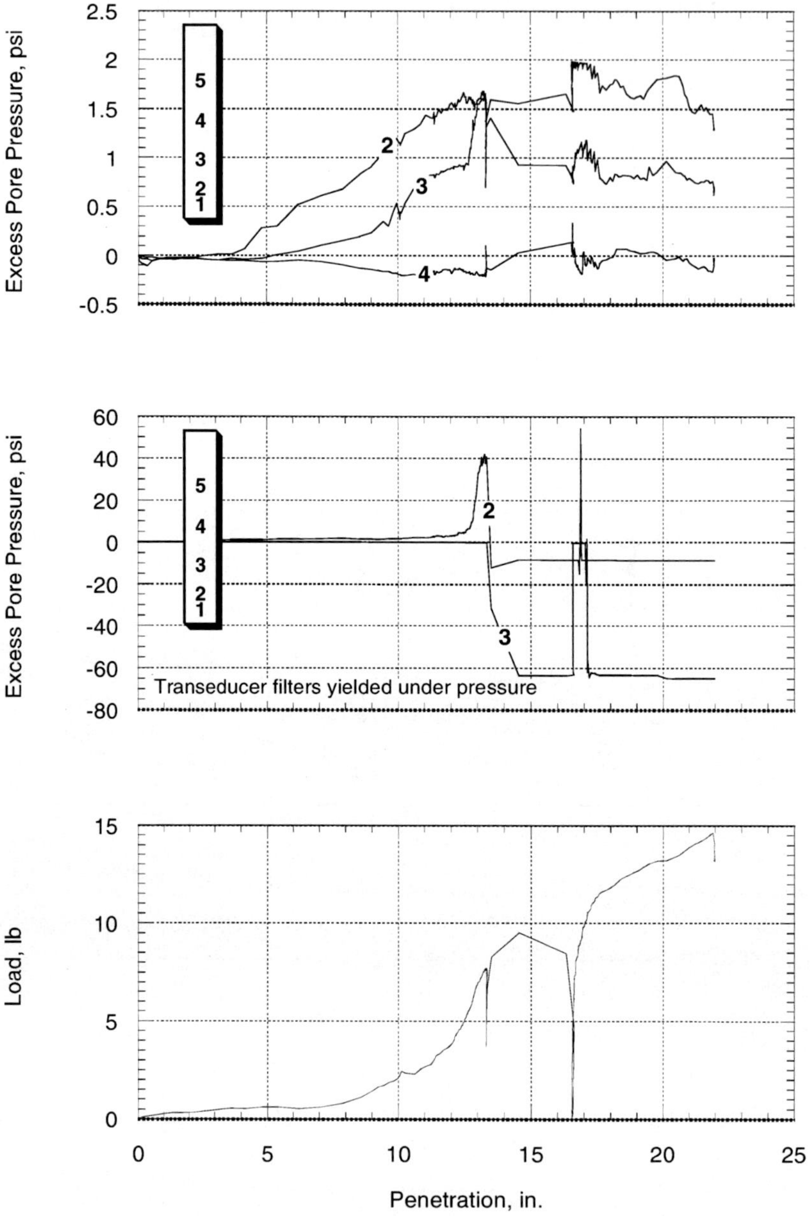

Fig. 9.23 Pore Pressure Distribution (a) Outside, (b) Inside a Pile (c) Pushed in Saturated Sand

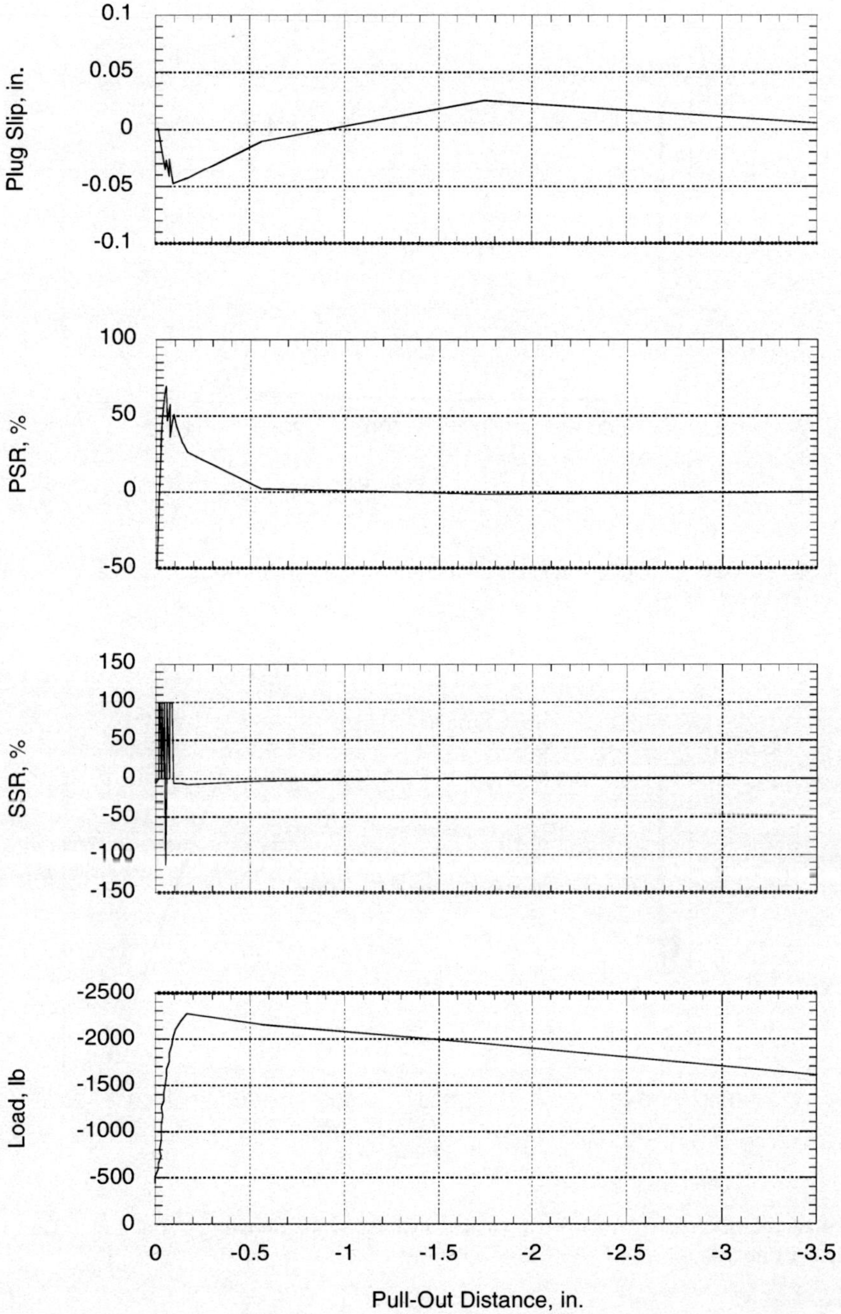

Fig. 9.24 Plug Slip, PSR, and SSR During Pull-Out Test in Saturated Sand

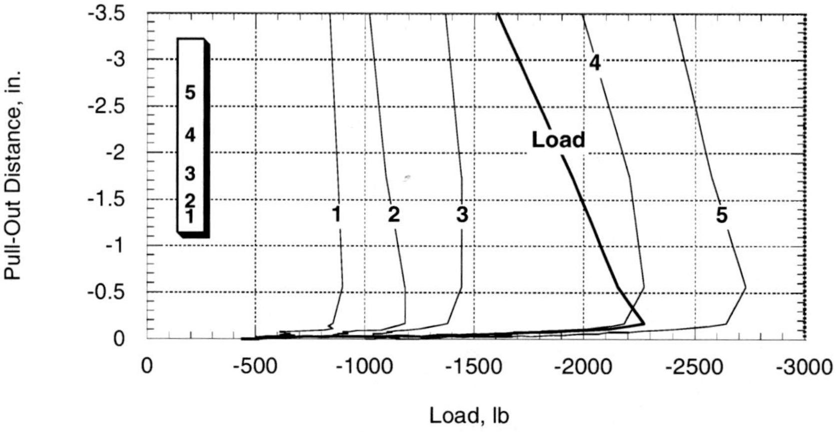

Fig. 9.25 Pull-Out Curve of a Pile, Pushed in Saturated Sand, Showing Axial Load Transfer Outside the Pile

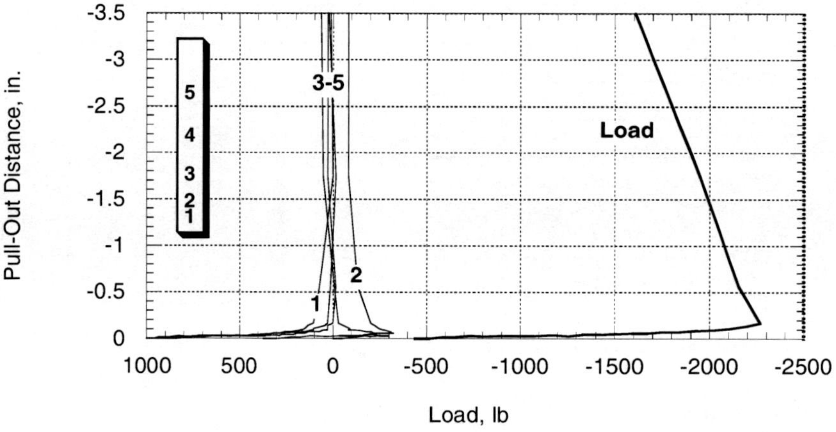

Fig. 9.26 Pull-Out Curve of a Pile, Pushed in Dry Sand, Showing Axial Load Transfer Inside the Pile

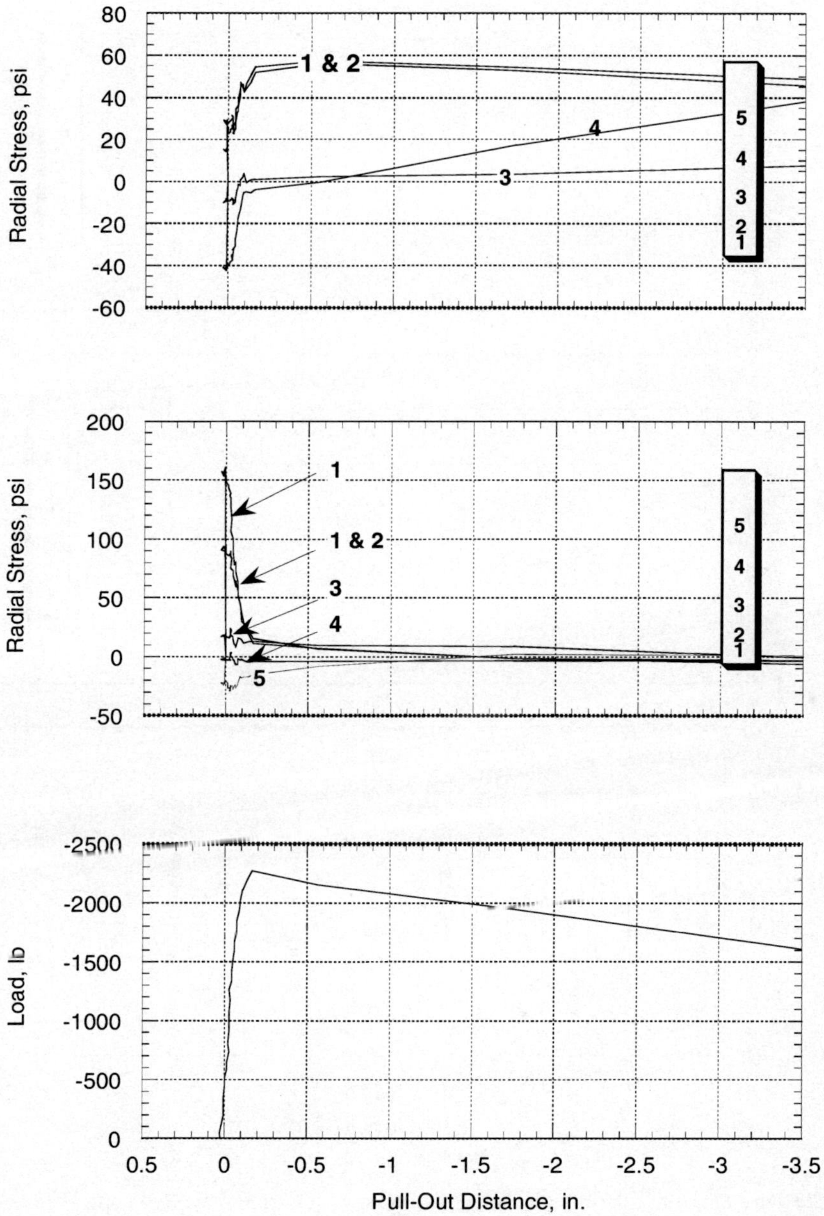

Fig. 9.27 Pressure Distribution during Pull-Out (a) Outside The Pile (Top), (b) Inside The Pile (Middle), and (c) Load-Displacement Curve (Bottom). (Pile Installed by Pushing in Saturated Sand)

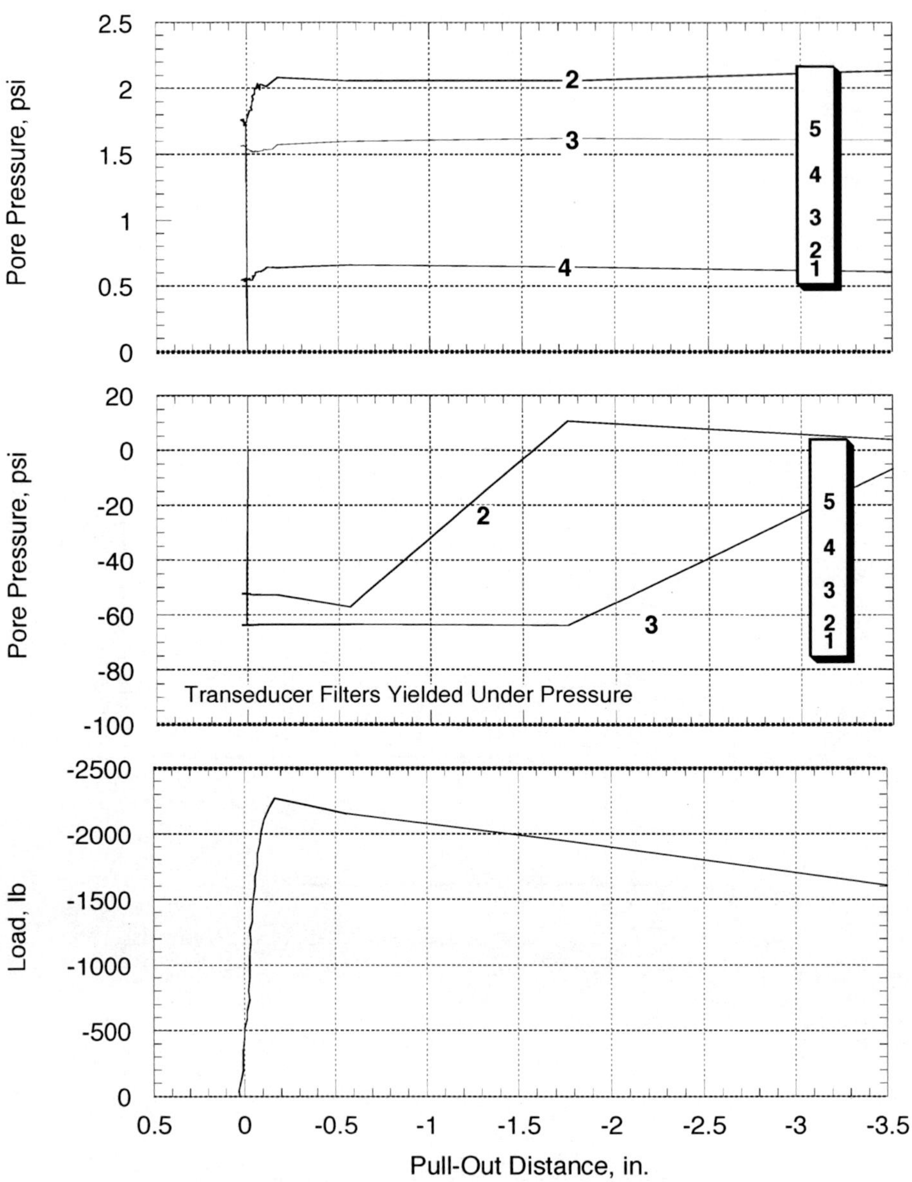

Fig. 9.28 Pore-Pressure distribution During Pull-Out (a) Outside the Pile (Top), (b) Inside The Pile (Middle), and (c) Load-Displacement Curve (Bottom). (Pile Installed by Pushing in Saturated Sand).

9.5.1 *Jacking (Pushing) Stage*

During the first 13 in. of pushing, the pile developed little skin friction, and less force was required for pushing than piles in dry sand (Fig. 9.14–9.16). Plugging occurred, abruptly, at a penetration of 13 in. and was readily identifiable by a sharp increase in the load transferred through the internal pile wall (Fig. 9.14), accompanied by a sharp decrease in the load transferred through the external pile wall (Fig. 9.15). After plugging, the force required to push the pile was larger than that in dry sand. Like piles pushed in dry sand, 80% of the end bearing was transferred to the internal pile wall in a length equal to one pile diameter (Fig. 9.15). The remaining 20% was transferred in the next diameter.

As the pile penetrated into the ground, the elevation of the soil in the pile increased and reached a maximum elevation 5 in. above the ground level outside the pile (Fig. 9.17). Plugging is indicated by the specific recovery ratio (SRR) approaching zero at a penetration of 13 in. Another indicator of plugging is the change in the slope of the PLR-penetration curve at a penetration of 13 in. After plugging the surface of the plug moved downward at the same rate of pile penetration.

9.5.2 *Load Test*

A plug stick-slip mechanism occurred in the middle of the load test stage and is indicated on Fig. 9.14–9.15 by sharp oscillations in the strain gage readings at an applied load in the range of 8–11 kips. This mechanism occurred in a number of tests in saturated sand, and could by identified during the test by the pile suddenly moving then stopping. A knocking sound in the hydraulic ram occurred when the plug developed a sudden increase in resistance, which stopped the ram. It is interesting to note that the sum of the strain gage readings in Fig. 9.16 does not show any oscillations, which indicates that during the stick-slip mechanism, the pile capacity cycled between the two pile walls, but the sum of the capacity at any given time was a constant.

The stick-slip mechanism occurred faster than data was sampled and accordingly applied loads in the range of 8-11 kips were not included in the axial load distribution curves (Fig. 9.18–9.19). Skin friction distributions were computed from axial load distributions and are shown in Fig. 9.20-9.21.

Outside the pile, no skin friction was developed in the first three diameters (Fig. 9.20). At deeper penetrations skin friction increased linearly with depth. The skin friction distribution is different from piles in dry sands in two ways. First, residual stresses did not affect the stress distribution, and negative skin friction did not develop. Second, the developed skin frictions are higher in saturated sand than in dry sand.

Inside the pile, large skin frictions developed near the tip. Skin frictions, apparently, reached a maximum two diameters up from the tip, and decreased sharply with further distance away from the tip. Reduction in the skin friction near the tip may have resulted from relaxation of the end bearing resistance.

High radial stresses were developed inside the pile, near the tip, and decreased with distance away from the tip (Fig. 9.22a). The stresses approached 270 psi, yet they were approximately 15% lower than the stresses developed in the pile pushed in dry sand. As discussed earlier these stresses must have resulted from arching of the sand due to plugging.

Outside the pile, the measured radial stresses appear to be higher at the center of the pile than at the tip or top (Fig. 9. 22b).

Pore pressures were measured outside the pile throughout installation and loading. A small build up of pore pressure occurred outside the pile. Excess pore pressures reached a maximum near the tip and reduced with distance away from the tip (Fig. 9.23). A maximum excess pore pressure of 2 psi was recorded 5 inches away from the tip.

Pile installation resulted in very large radial stresses, particularly inside the pile. These stresses caused the filters of some pore pressure transducers to yield under pressure, and touch the sensing face. Other filters were stripped off the transducers. The pore pressure transducers were designed to have a pore pressure rating of 25 psi, with a factor of safety of 2. In one case, the transducer which is located inside the pile nearest to the tip was subjected to a radial stress in excess of 300 psi and its filter and sensing face were both fused and yielded together. It was possible to retrofit the transducers mounted on the outside of the pile with external filters. The retrofit was successful except for the transducer located nearest to the tip, which worked intermittently. The transducers located inside the pile were not accessible. However, the two transducers located 5 and 10 in. away from the tip continued to work intermittently, depending on the applied radial stress.

9.5.3 Pull-Out Test

The pile's pull-out capacity represented 20% of the capacity in compression (Fig. 9.24). The plug was retained inside the pile during pull-out, and essentially no plug slippage occurred.

The strain gages appear to have experienced large zero shifts, and their absolute values are not reliable (Fig. 25). However, strain gage readings indicate that the capacity was developed entirely from friction outside the pile (Fig 9.25–9.26). Like piles in dry sand, the plug contained large residual stresses which were released within 0.1 in. of pull-out (Fig. 9.26). The radial stress distribution inside the pile also indicates that the plug stresses were relieved (reduced) within 0.1 in of pull-out.(Fig 9.27).

The pore pressures measured outside the pile during pull-out remained unchanged (Fig 9.28). Pore pressure measurements inside the pile were unreliable.

9.6 Load Tests on the Pile Driven in Saturated Sand

Measurements taken during driving, load testing, and pull-out of the pile in saturated sand are presented in Fig. 9.29–9.34, Fig 9.35–9.44, and Fig. 9.45–9.49, respectively.

9.6.1 Driving Stage

Open ended piles in saturated sand typically drove much easier than closed ended piles in dry sand. The driven pile discussed in this section encountered easy driving, and approximately 90 blows were required for installation.

As the pile penetrated into the ground, the elevation of the soil core increased at a rate which was approximately half of that of unplugged pushed piles (Fig. 9.29). However, unlike pushed piles, a plug did not form at any stage of the driving process.

The response of the pile-soil system during four consecutive blows in the middle of driving is shown in Fig. 9.30. The set per blow was approximately 0.15 in. The frequency response of the position transducer used to monitor plug movements was slower than desired. However, it is clear that the plug continued to incrementally rise inside the pile after each blow.

A surge in pore pressure of approximately 0.4 psi was also recorded concurrent to each blow (Fig 9.30). The rise time of the pore pressure wave was approximately 0.1 sec. The pressure surge dissipated 0.1 sec later[3]. Surges of pore pressure resulted in a small build up of pore pressure outside the pile (Fig. 9.31).

Excess pore pressures reached a maximum near the tip and reduced with distance away from the tip. A maximum excess pore pressure of 1.25 psi was recorded 5 inches away from the tip, which represent 6.5% of the initial in situ effective stress. A build up of pore pressure was also recorded inside the plug (Fig 9.32). The excess pore pressures inside the plug were uniform throughout installation. The excess pore pressure in the plug was 1.5 psi, which represents 7.5% of the initial in situ effective stress.

[3] A continuous sampling frequency of 10 Hz was used to record during driving, which requires that the hammer impact and the recording trigger be synchronized in order to observe surges of pore pressure. Synchronization was not performed and accordingly the observed behavior could be observed only intermittently on the driving record. It is possible to use periodic sampling and sample at a higher frequency.

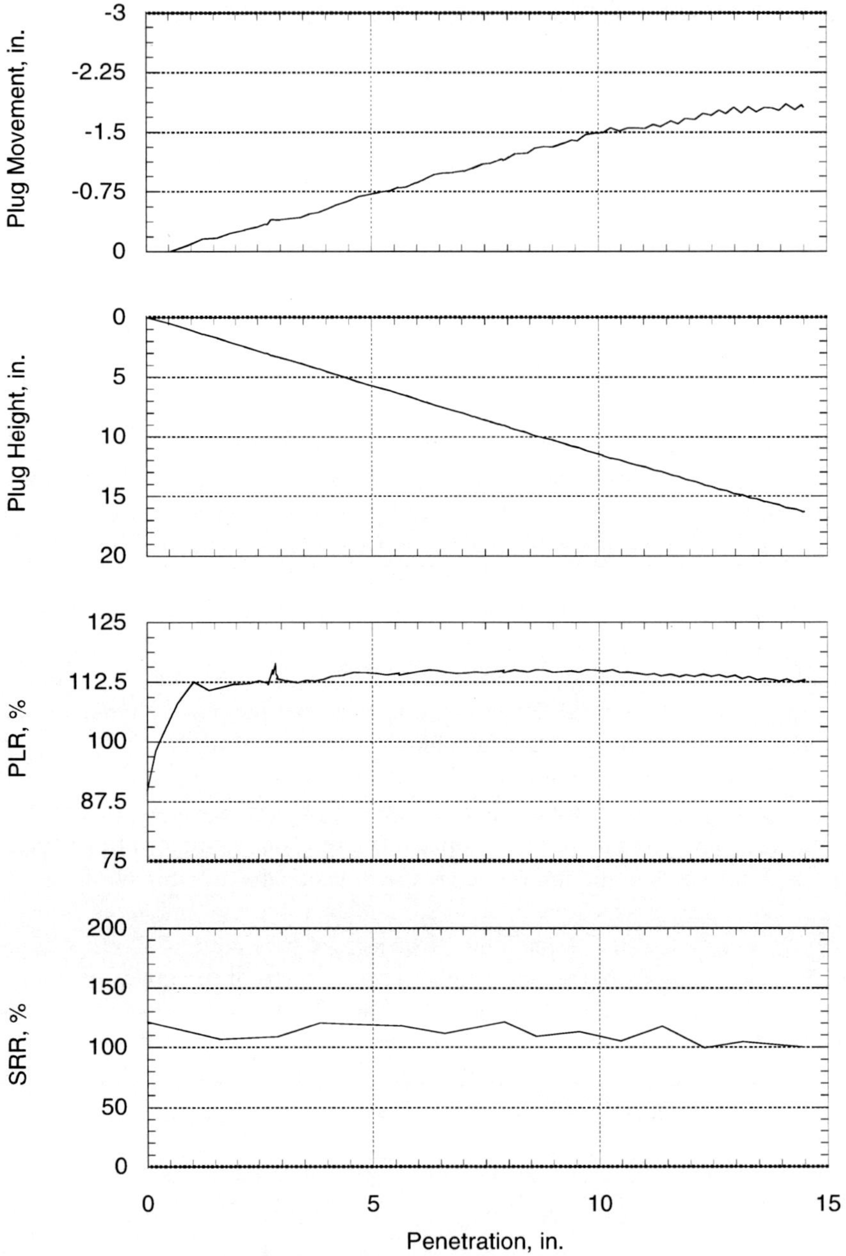

Fig. 9.29 Plug Movement, SRR, and PLR during Driving

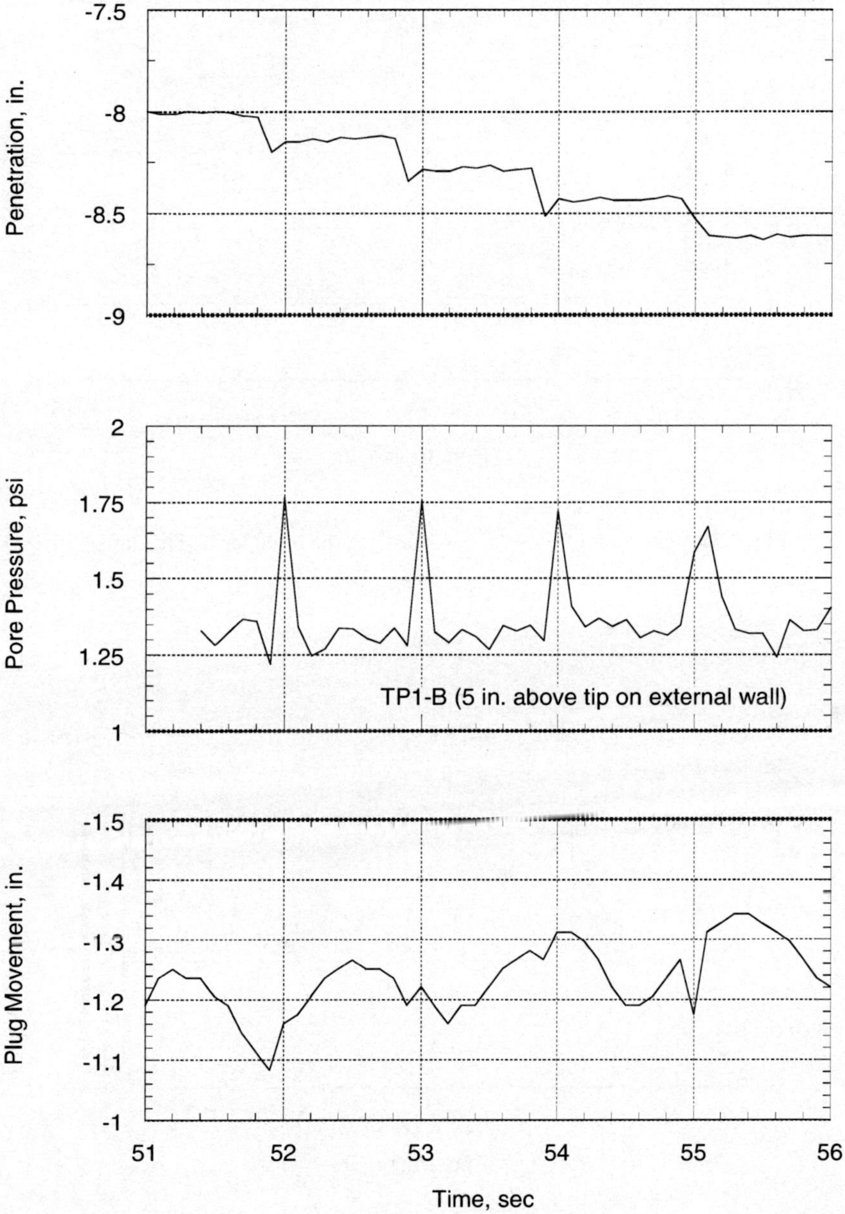

Fig. 9.30 Penetration, Pore Pressure, and Plug Response per Hammer Blow

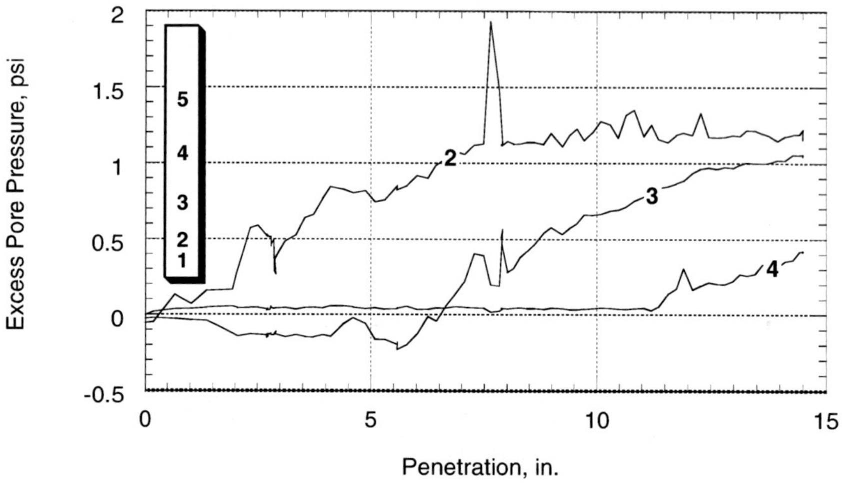

Fig. 9.31 Build-up of Pore Pressures Outside the Pile During Driving

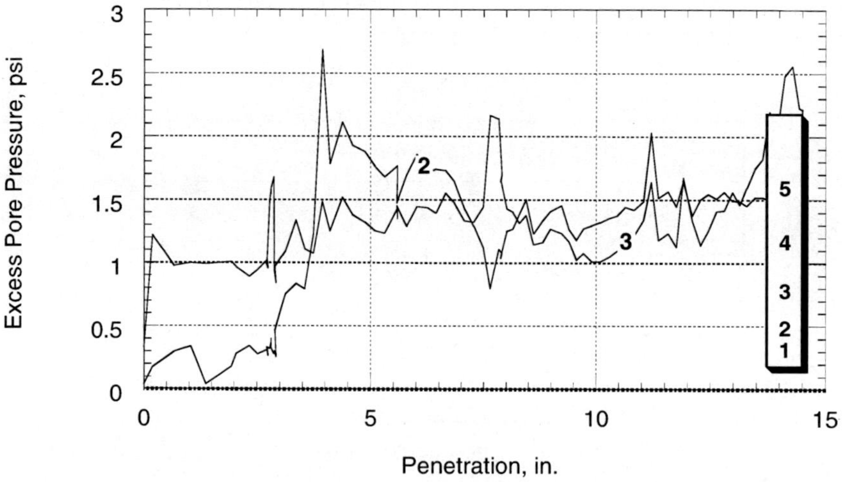

Fig. 9.32 Build-up of Pore Pressures Inside the Pile During Driving

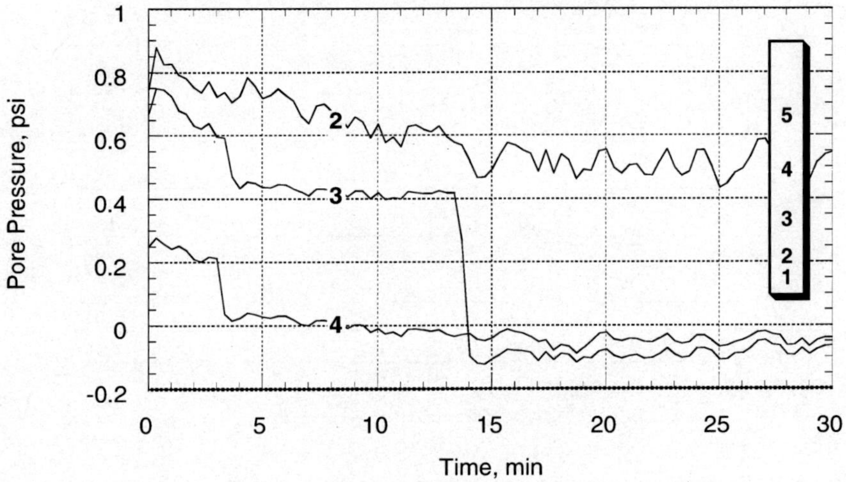

Fig. 9.33 Dissipation of Pore Pressures Outside the Pile after Driving

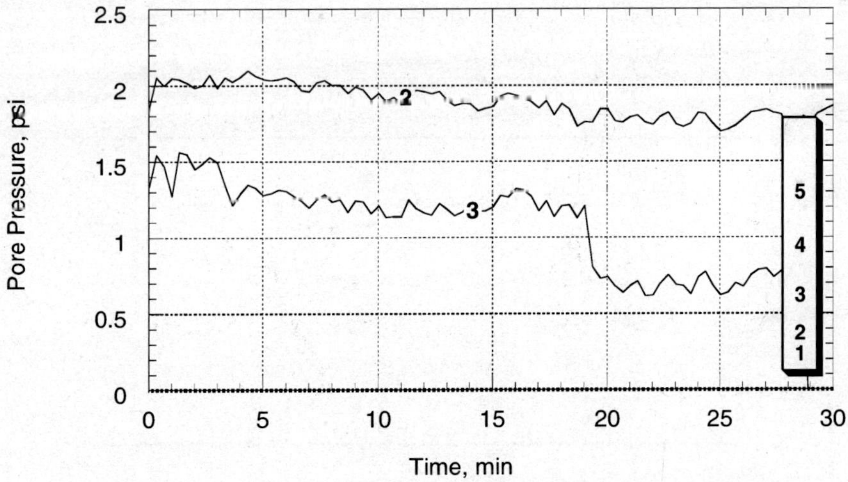

Fig. 9.34 Dissipation of Pore Pressures Inside the Pile after Driving

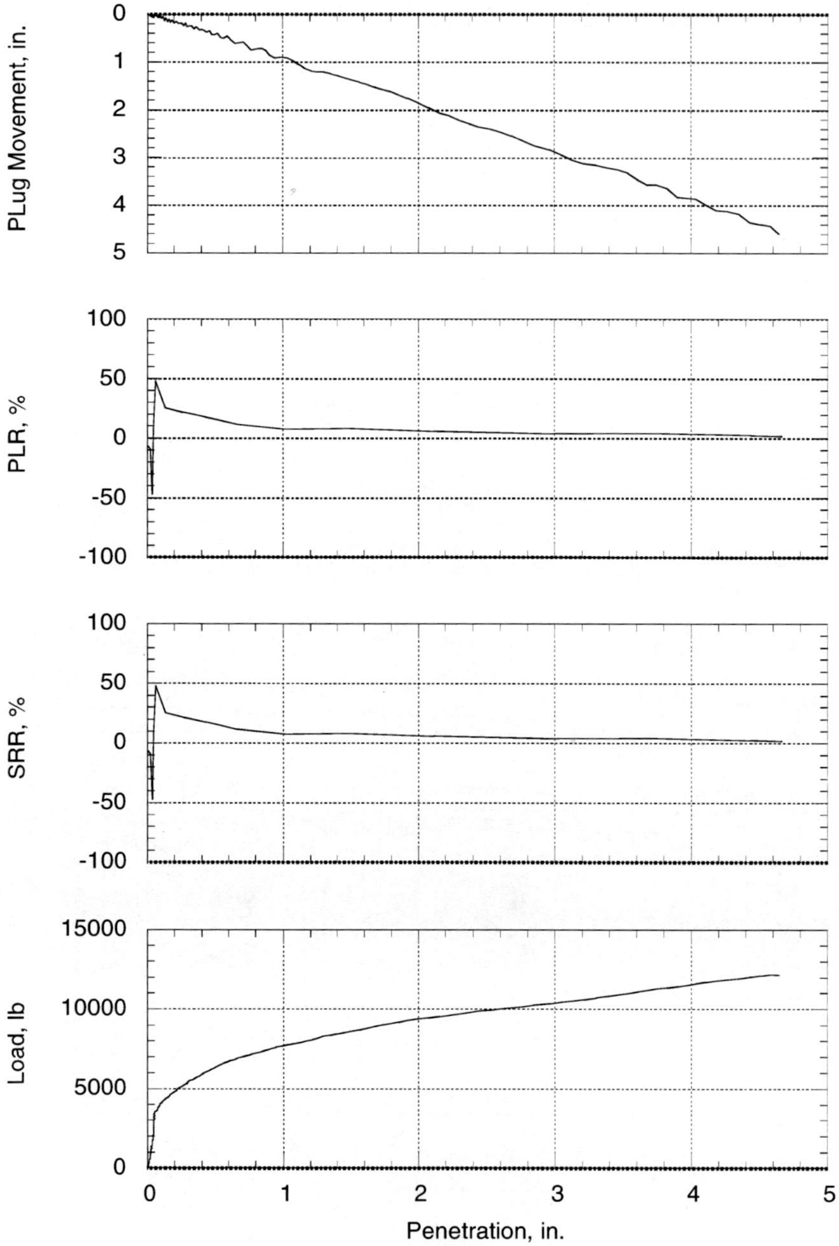

Fig. 9.35 Plug Movement, SRR, and PLR during Load Test

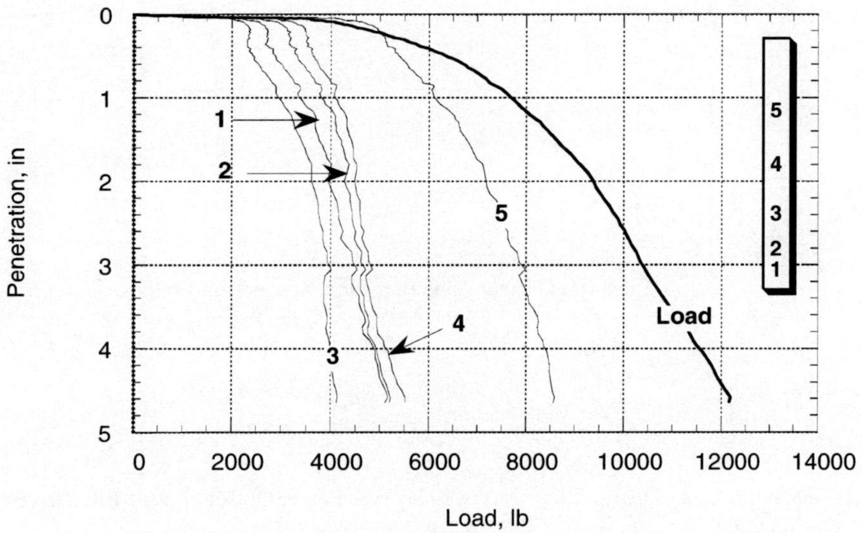

Fig. 9.36 Load-Settlement Curve of a Pipe Pile, Driven in Saturated Sand, Showing Axial Load Transfer Outside the Pile

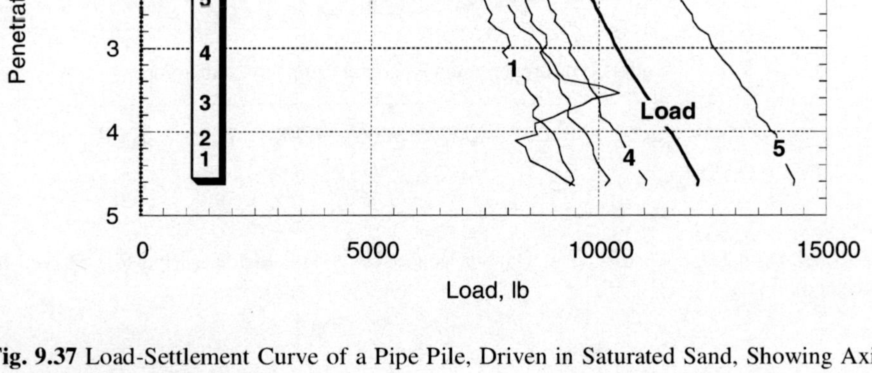

Fig. 9.37 Load-Settlement Curve of a Pipe Pile, Driven in Saturated Sand, Showing Axial Load Transfer Inside the Pile

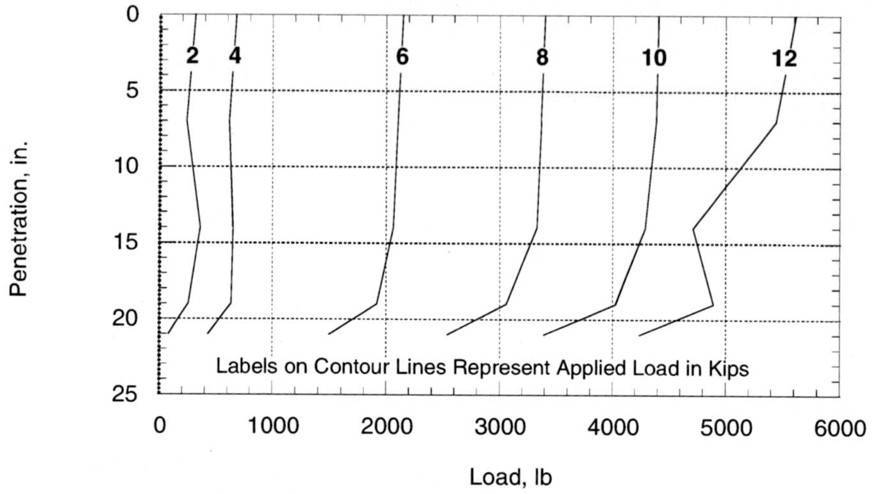

Fig. 9.38 Axial Load Distribution, During Load Test Stage, Outside a Pipe Pile Driven in Saturated Sand

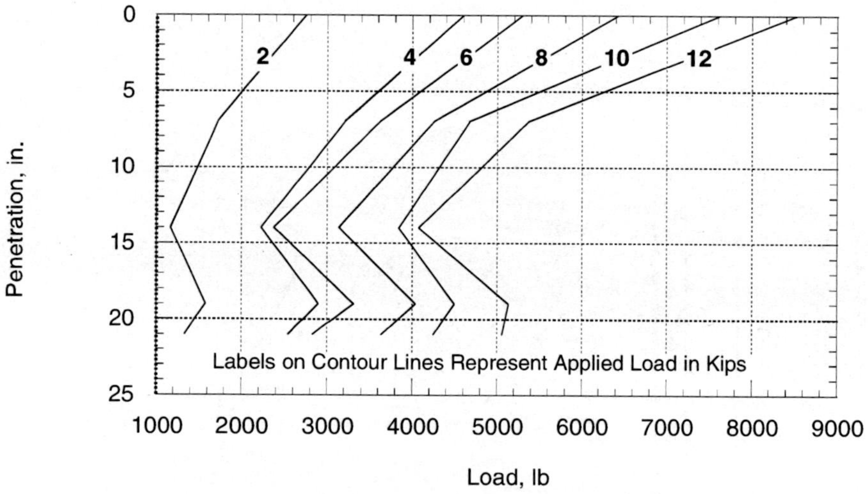

Fig. 9.39 Axial Load Distribution, During Load Test Stage, Inside a Pipe Pile Driven in Saturated Sand

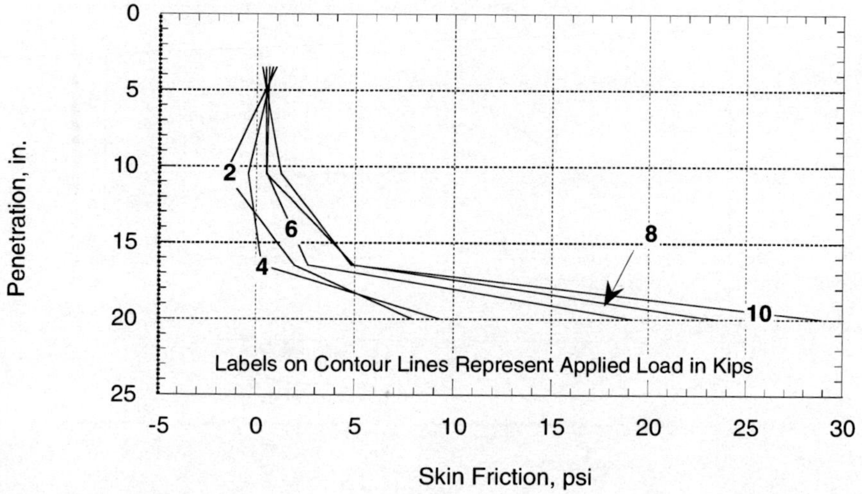

Fig. 9.40 Skin Friction Distribution, During Load Test Stage, Outside a Pile Driven in Saturated Sand

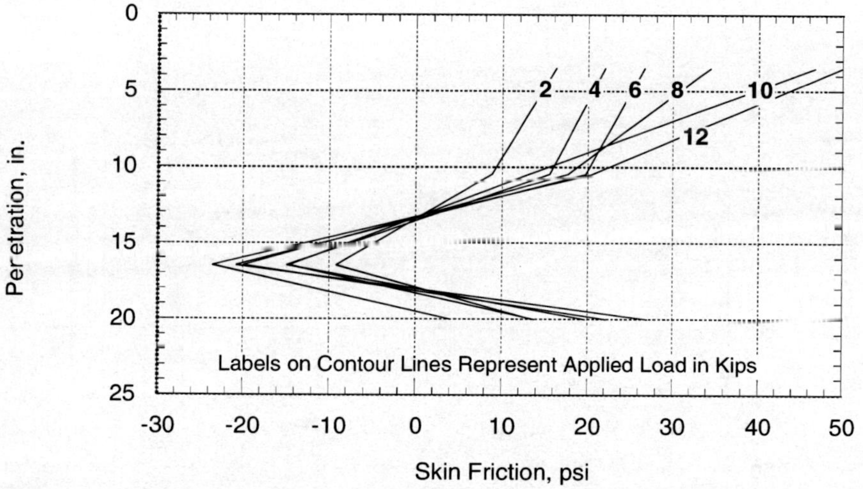

Fig. 9.41 Skin Friction Distribution, During Load Test Stage, Inside a Pile Driven in Saturated Sand

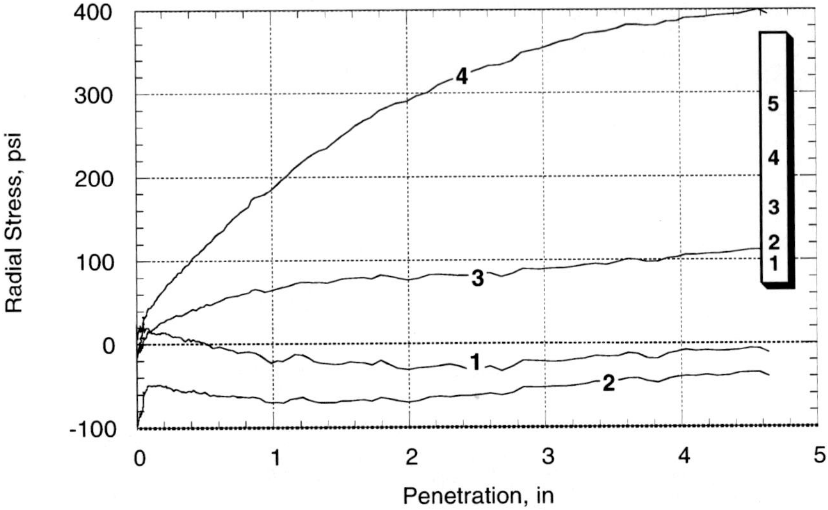

Fig. 9.42 Radial Stresses Acting, During Load Test Stage, Outside a Pipe Pile Driven in Saturated Sand

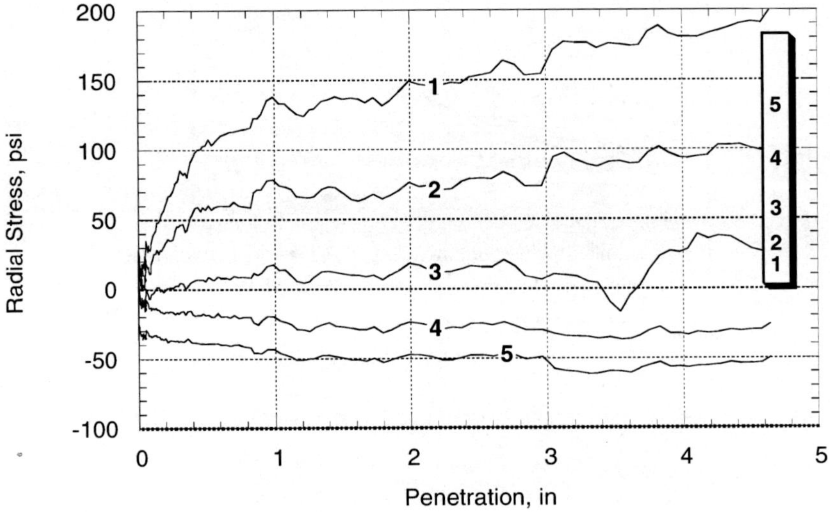

Fig. 9.43 Radial Stresses Acting, During Load Test Stage, Inside a Pipe Pile Driven in Saturated Sand

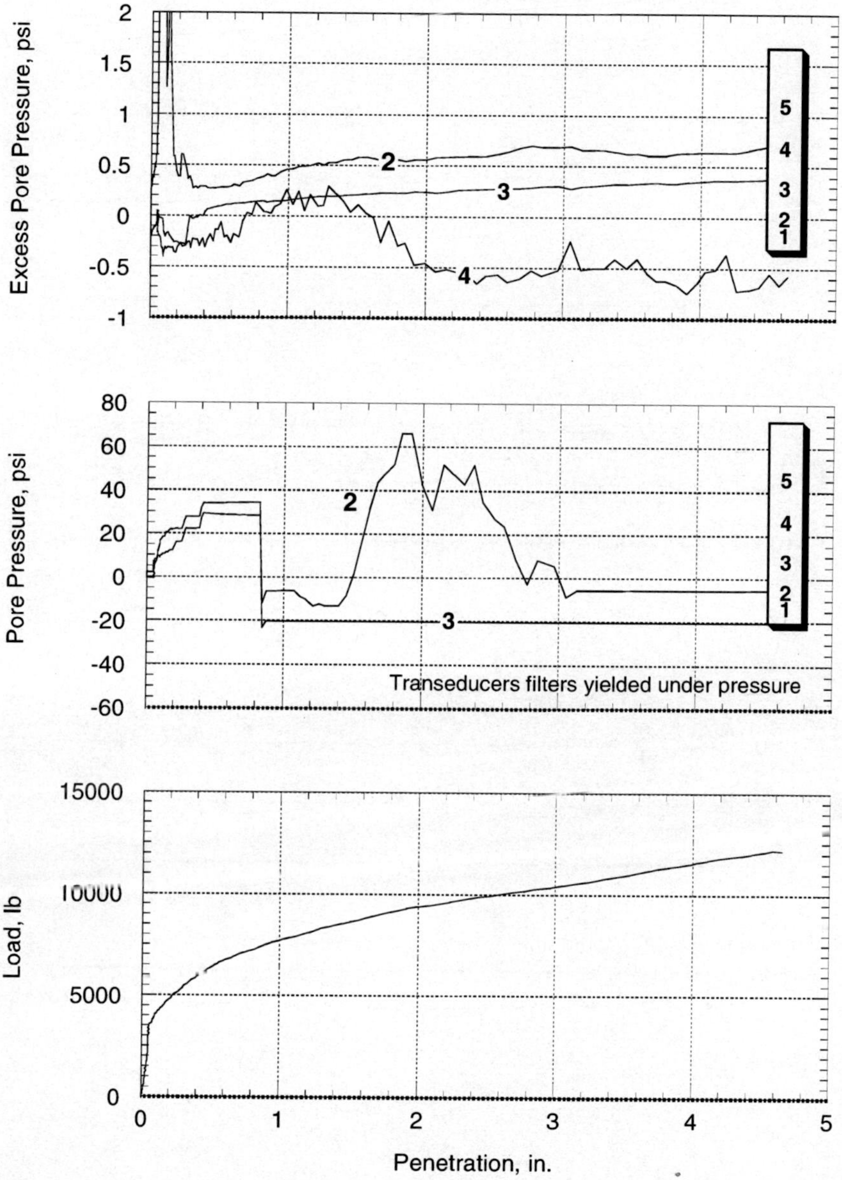

Fig. 9.44 Pore Pressure Distribution (a) Outside, and (b) Inside a Driven Pipe During (c) a Load Test

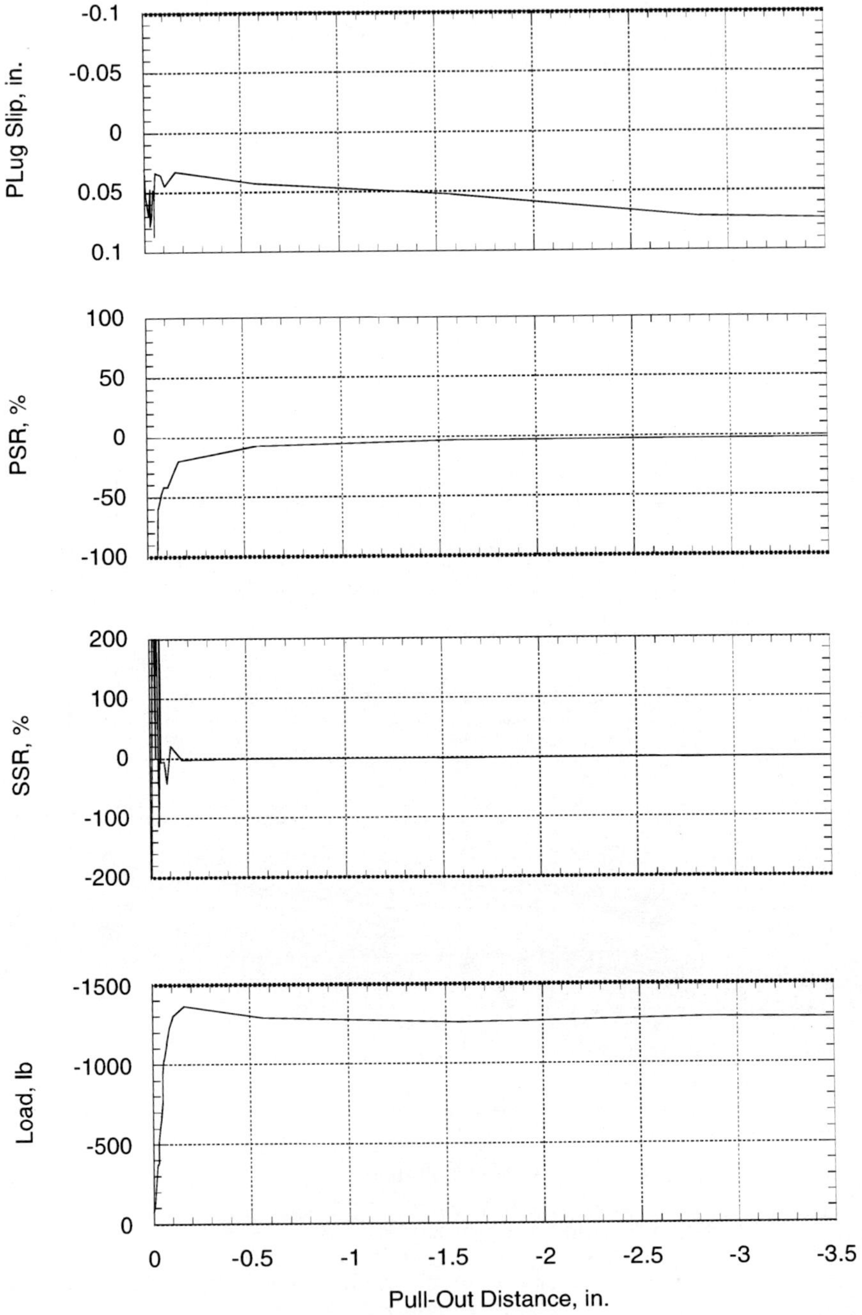

Fig. 9.45 Plug Slip, PSR, and SSR during Pull-Out Test

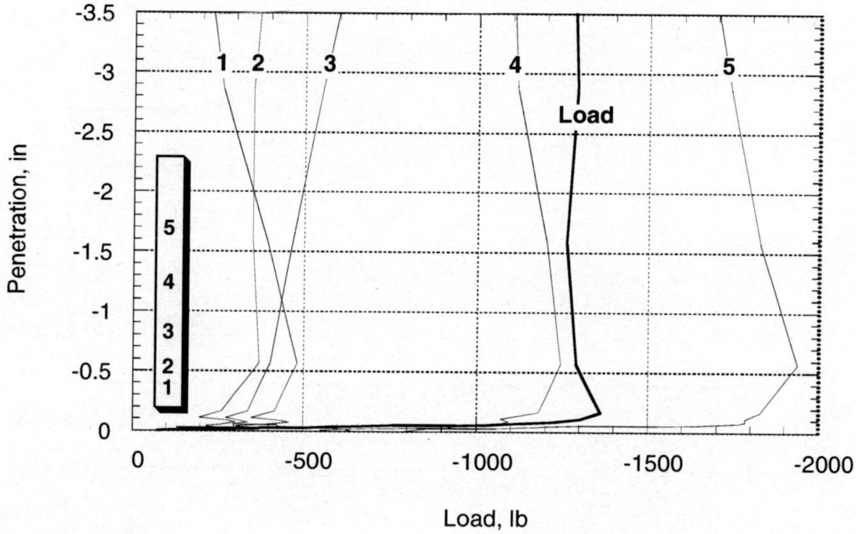

Fig. 9.46 Pull-Out Curve of a Pile, Driven in Saturated Sand, Showing Axial Load Transfer Outside the Pile

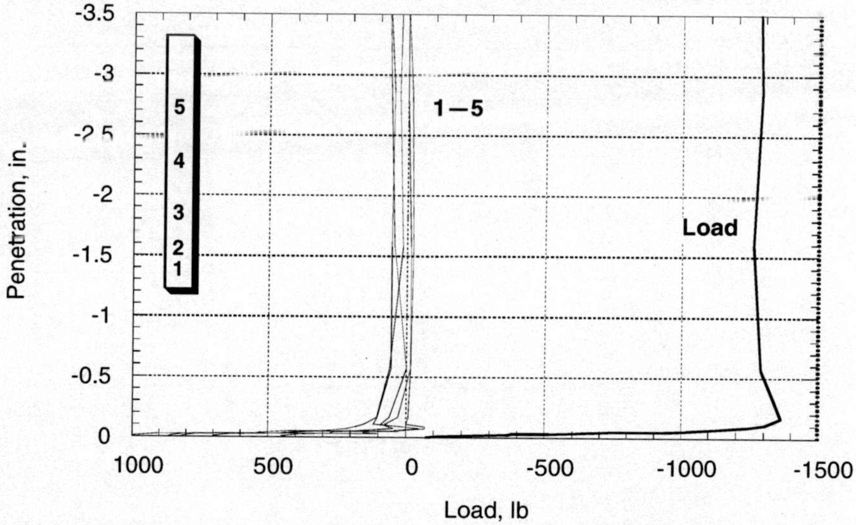

Fig. 9.47 Pull-Out Curve of a Pile, Driven in Saturated Sand, Showing Axial Load Transfer Inside the Pile

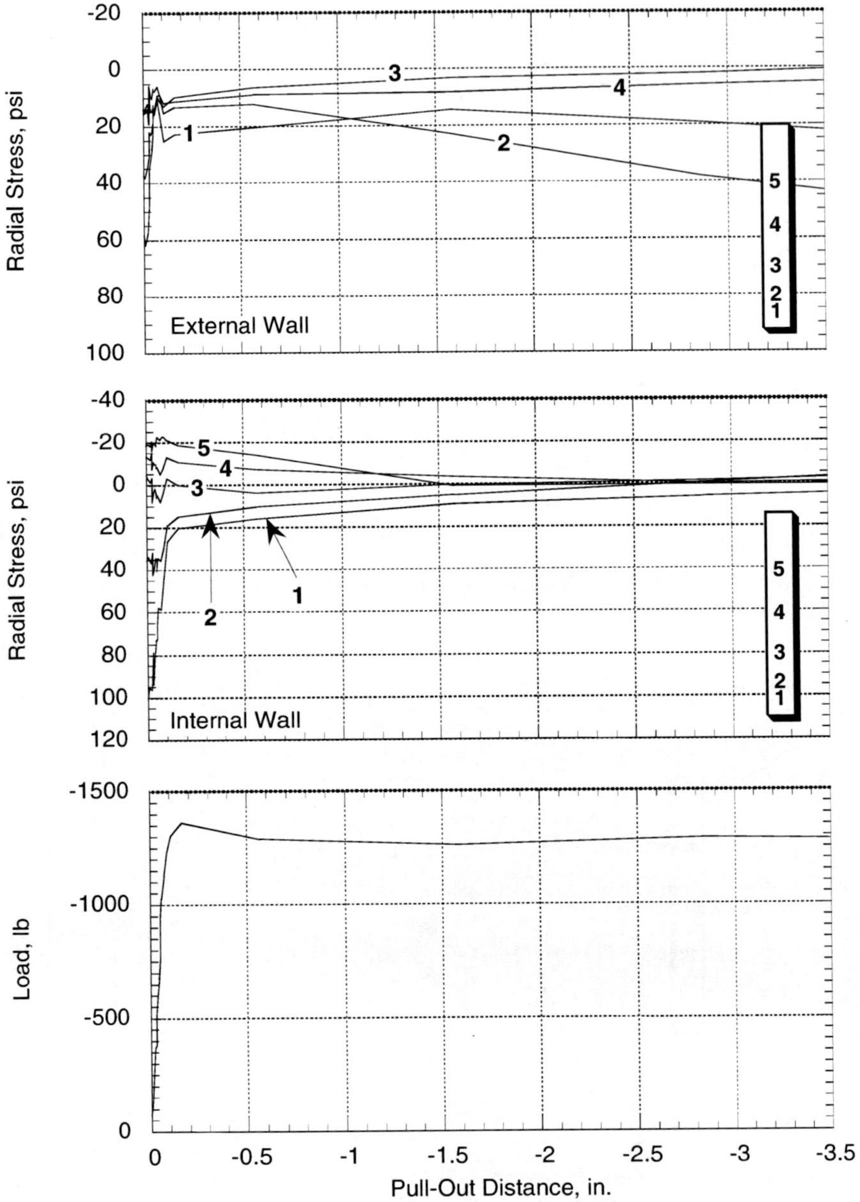

Fig. 9.48 Radial Stress Acting on (a) The External (Top) and (b) Internal Surfaces (Middle) of a Pile Driven in Saturated Sand During (c) Pull-Out Test (Bottom)

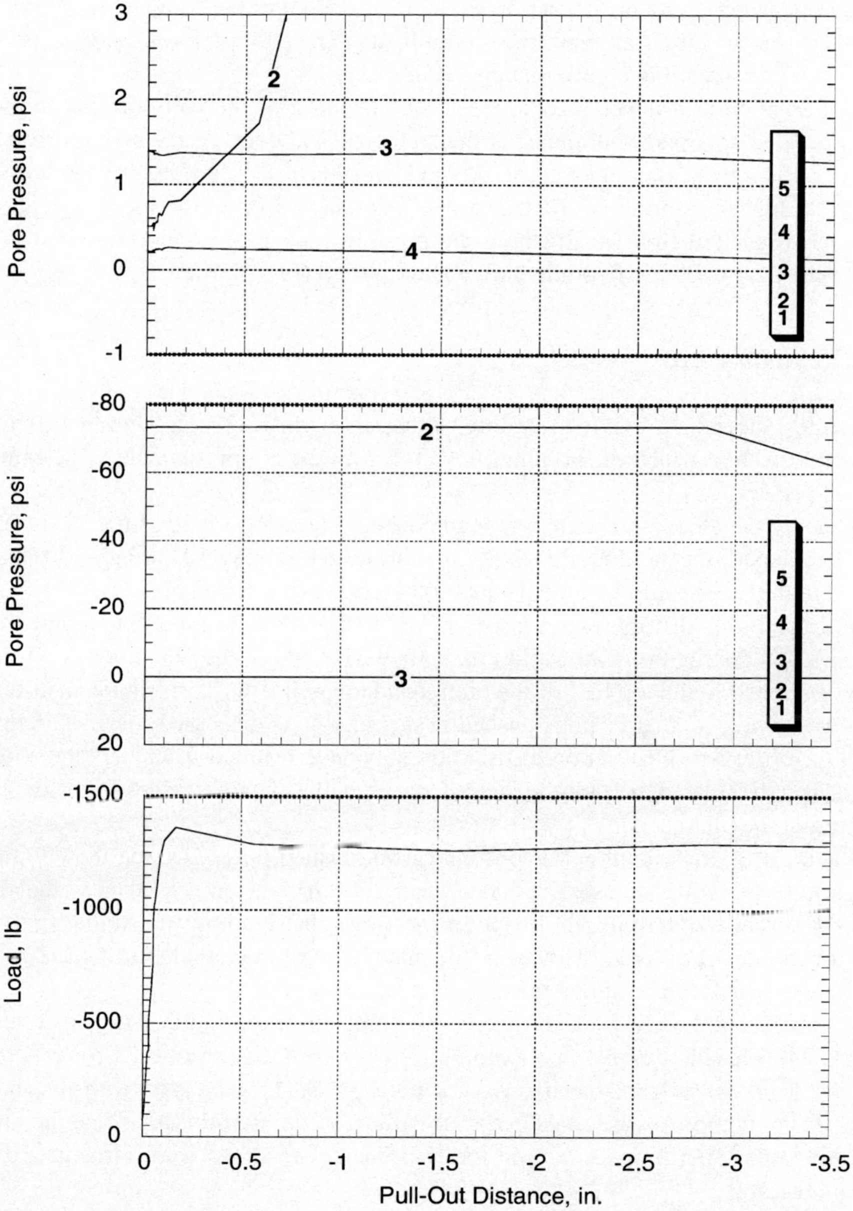

Fig. 9.49 Pore Pressures Acting on (a) The External (Top) and (b) Internal Surfaces (Middle) of a Pile Driven In Saturated Sand During (c) Pull-Out Test (Bottom)

The sand in the pressure chamber was covered by a freely draining geocomposite filter. After driving, excess pore pressures, outside the pile, dissipated in times which depended on their distance from the top geocomposite drainage filter, with further distances requiring more time to dissipate (Fig 9.33). All excess pore pressures were dissipated in approximately 20 minutes.

The position transducer used to track the elevation of the soil core inside the pile provided only partial drainage at the top of the soil core. At the end of driving when the hammer was removed, it was evident that the pore pressure inside the plug was higher than the hydrostatic value. A sudden drop in the plug's pressure may have resulted from re-attaching the position transducer, which required the top of the plug to be freely drained for a short period (Fig. 9.34).

9.6.2 Load Test

The pile plugged during load testing, although it did not plug during driving (Fig 4.35). The surface of the plug moved downward at approximately the same rate of penetration.

The pile developed less capacity than pushed piles. Relatively, little skin friction developed outside the pile (Fig 9.36). Inside the pile, end bearing was transferred to the inner wall in a length equal to two pile diameters (Fig 9.37).

The axial load distributions shown in Fig. 9.38–9.39 were used to compute the skin friction distributions shown in Fig. 9.40–9.41. Outside the pile, little skin friction developed in the top half of the pile. Inside the pile, the skin friction distribution was similar to that of piles pushed in dry sand. As discussed for piles in dry sand the observed skin friction distribution may have resulted from a combination of a large stress gradient and the large distance between strain gages in the upper portion of the pile.

Radial stress distribution was different from pushed piles. Outside the pile the magnitude of radial stresses increased near the top and decreased near the tip (Fig. 4.42). Inside the pile, the shape of the stress distribution was similar to that of pushed piles (Fig 4.43). However, the magnitude of the radial stresses was half of that observed for pushed piles.

A small build up of excess pore pressures occurred during the load test (Fig 9.44). Outside the pile, a maximum excess pore pressure of 0.75 psi was recorded 5 in. away from the tip. Excess pore pressures decreased with distance away from the tip. As discussed earlier the filters of the transducers inside the pile yielded under lateral pressure, and the transducers' readings were influenced by earth pressures.

9.6.3 *Pull-Out Test*

The pile's pull-out capacity represented 20% of the capacity in compression (Fig. 9.45). The plug was retained inside the pile during pull-out, and essentially no plug slippage occurred.

Strain gage readings indicate that the capacity was developed entirely from friction outside the pile (Fig 9.46–9.47). The strain gages appear to have experienced a large zero shift, and their absolute values are not reliable (Fig. 9.46). It is clear however that, like pushed piles, the plug contained large residual stress which were released within 0.1 in. of pull-out (Fig. 9.26). The radial pressure distribution inside the pile also indicates that the plug stresses were released (reduced) within 0.1 in of pull-out (Fig 9.48).

The pore pressures measured outside the pile during pull-out remained relatively unchanged (Fig 9.28). As discussed earlier, pore pressure measurements inside the pile were unreliable.

9.7 Observed Behavior of Piles in Sand

The results of the testing program provide a number of important insights into the behavior of pipe piles during installation and loading, as follows:

9.7.1 *Pile Plugging*

Piles plug during pushing, but not during driving. Formation of a plug during installation by pushing did not significantly change the penetration resistance. When a plug formed, 80% of the end bearing was transferred to the internal surface of the pile in a length equal to (or less than) one pile diameter from the tip. The remaining end bearing is transferred in a the second pile diameter.

Plugging results in the formation of very large radial stresses inside the pile. These stresses were larger for pushed piles than for driven piles. The stresses were also larger for piles in dry sand than saturated sand.

During pull-out, very small pile movements were required to release the plug stresses. The plug is retained in piles installed in saturated sand and slipped in piles installed in dry sand.

9.7.2 Buildup of Pore Pressures

Pile driving resulted in a pore pressure surge near the tip. The rise and decay times of the pressure surge were each approximately 0.1 sec long. The pressure surge resulted in a small build up of pore pressure during driving. The measured pore pressures were approximately 7.5% of the initial effective stresses. Although it was not possible to continuously record pore pressure inside the plug, due to equipment failures, evidence exists that the pore pressure inside the plug was higher than the hydrostatic value.

The measured increase in excess pore pressure during driving is not sufficient to significantly influence plugging. Nevertheless, build up of pore pressure due to cyclic loading depend on the number of loading cycles. Only 90 hammer blows were required to install the pile. In the field, larger pore pressures may develop during driving because a much larger number of hammer blows is used.

A small build up of pore pressure was measured during installation by pushing and during load tests. The measured pore pressure ranged between 4% and 10% of the initial in situ effective stresses.

Pore pressures dissipated within 20 minutes from installation. According to theory of radial consolidation, dissipation of pore pressures depends on the pile diameter and may take a longer duration for production size piles.

9.7.3 Load Transfer

For piles pushed in dry sand, skin friction was greatly influenced by residual stresses. The effects of residual stresses were prominent until the load used to install the pile by pushing was exceeded. Residual stresses appear to have lesser effects on piles installed in saturated sand.

Load transfer outside the pile appeared to increase with depth despite the fact that the pressure chamber had initial isotropic stresses.

9.8 Conclusions

The results of the load tests performed using the instrumented pile provide convincing evidence of the value of the double-wall concept in pile foundation research. The pile performed as designed and provided rare data on soil pile interaction. The full potential of the double wall concept will be realized when it is implemented in a production-size pile.

The experimental results of this study contribute to a better understanding of pile behavior, but they also raise additional questions. Research should continue to study the various aspects of soil structure interaction. In particular, the following issues should be examined in detail:

- The effect of residual stresses on the development of negative skin frictions inside and outside the pile
- Build-up of pore pressures during installation and subsequent loading, and its effect on pile behavior.
- The effects of pile plugging (or lack there of) during installation on the mobilized axial capacity during loading.

References

Alansari, A.: Capacity and Behavior of Steel Pipe Piles in Dry Sand. Ph.D. Dissertation, The University of Texas at Austin (1999)

Al-Shafei, K.A., Cox, W.R., Helfrich, S.C.: Pile Load Tests in Dense Sand - Analysis of Static Test Results. In: Proc. Offshore Technology Conf., pp. 83–103 (1994)

Hvorslev, M.J.: Subsurface Exploration and Sampling of Soils for Civil Engineering Purposes, ASCE Research Project of the Committee on Sampling and Foundations, ASCE (1949)

Iskander, M.: An Experimental Facility to Model the Behavior of Steel Pipe Piles in Sand. Ph.D. Dissertation, The University of Texas at Austin (1995)

Malhotra, S.: Effect of Wall Thickness on Plugging of Open Ended Steel Pipe Piles in Sand. In: Contemporary Issues in Deep Foundations, GSP 158 (2007), doi:10.1061/40902(221)36

Paikowsky, S.G.: A Static Evaluation of Soil Plug Behavior with Application to the Pile Plugging Problem, Ph.D. Thesis, Massachusetts Institute of Technology (1989)

Robinsky, E.I., Morrison, C.F.: Effect of Shape and Volume on The Capacity of Model Piles in Sand. Canadian Geotechnical Journal 1(4), 189–204 (1964)

Vesic, A.S.: A Study of Bearing Capacity of Deep Foundations, Final Report on Project B-189, Georgia Institute of Technology, Atlanta, Georgia (1967)

10. Summary and Conclusions

Abstract. One of the major difficulties in predicting the axial capacity of pipe piles in sand has been a lack of understanding of the physical processes which control the behavior of piles during installation and loading. The objectives of this research are to develop the experimental facilities necessary to identify these processes and perform load tests to study the phenomenon of pile plugging. A summary of the work performed, conclusions, and recommendations for future research are presented in this chapter.

10.1 Background

The research presented in this monograph has been motivated by the challenges facing the designers of offshore structures. Nevertheless, the research findings apply equally well to all pile foundations. Most offshore structures are supported on steel pipe piles which are designed in accord with *API Recommended Practice for Planning, Designing, and Constructing Fixed Offshore Platforms, RP-2A* (API, 1993). The guidelines for calculating the axial capacity of piles in sand have long been the source of considerable debate due to the geotechnical profession's inadequate understanding of pile behavior.

The API method uses a simple theoretical model which is calibrated using empirical load test data. The primary inadequacy of the API method is the data base used to calibrate the design method. At the present time, the data base has fewer load tests than the number of degrees of freedom of the pile-soil system, which precludes determining a unique set of calibration factors. Furthermore, the data base consists of piles that are generally shorter in length, smaller in diameter, and have less capacity than the piles being designed for offshore structures.

Analytical methods, alone, can not resolve the uncertainties in axial capacity predictions due to the complex nature of pile behavior. Experimental research on the physical processes that control the behavior of piles during installation and loading is required in order to resolve the issues involved. Considering the large number of variables involved in axial pile capacity development, as well as the high cost of load tests on full-scale instrumented piles, it is unlikely that a sufficient number of full-scale load tests can be performed to resolve the problem. Furthermore, the test conditions of full-scale load tests, generally, can not be controlled satisfactorily to isolate independent variables. Experimental research on model piles under controlled conditions can, however, be used to resolve some of the important design uncertainties.

M. Iskander: Behavior of Pipe Piles in Sand, SSGG, pp. 245–249.
springerlink.com

One of the objectives of this study is to develop a body of knowledge which will aid the designers of both offshore and terrestrial piles to obtain more precise capacity predictions.

10.2 Summary

Analysis of known case histories, in chapter two, indicates that a large scatter between measured and predicted pile capacities exists. Variations between measured and predicted pile capacities arise due to a number of factors such as severe changes in soil fabric and state of stress caused by pile driving, difficulties in measuring field soil properties, particularly in the marine environment, variations in loading details and pile installation procedures, behavior of the soil core (plug) during driving and subsequent loading, and soil structure interaction.

A close examination of the effects of the installation process on the capacity of piles, in chapter three, revealed that pile plugging has a considerable effect on axial pile capacity. An experimental program to study the relationship of plugging, generation of pore pressures, and redistribution of stresses during installation and loading was therefore initiated.

The experimental facility described in chapters four, five, and six was built to study the behavior of pipe piles in sand. This study is different from previous experimental studies in a number of ways. First, a double walled pile was used to isolate the frictional stresses acting against the external and internal surfaces of the pile wall. Second, a fast, automatic laboratory pile hammer capable of representing the phenomena which occur during pile driving was developed and used. Third, a data acquisition system capable of collecting data at a large scale was built. Fourth, this study is the first to provide continuous measurements of the elevation of the soil inside the pile during installation and loading. Fifth, a pressure chamber, feedback control system, loading frame, sand handling, pluviating, saturating, and drying apparatus have been integrated to allow convenient load testing of piles under simulated field conditions.

A review of the modeling capabilities of the developed facilities, in chapter 8, demonstrates that driving of typical offshore piles can be adequately represented using the developed facilities.

A testing program to investigate plugging of piles during installation and loading is presented in chapter 9. The results of the testing program validate the concepts used to design the test facility and provide important insights into soil behavior which are summarized next.

10.3 Conclusions

A number of conclusions may be drawn regarding the design and performance of the developed apparatus as well as regarding the behavior of piles in sand.

10.3.1 Conclusions Regarding Experimental Apparatus

10.3.1.1 The Double-Wall Pile

The results of the load tests performed using the instrumented pile provide convincing evidence of the value of the double wall concept in pile foundation research. The pile performed as designed, and provided data that has never been measured in the past. The full potential of the double-wall concept will be realized when it is implemented in a production-size pile.

The methods presented in chapter five were successful in protecting the pile instrumentation from failure during pile driving. The radial stresses acting inside the pile were approximately one order of magnitude larger than the free field stresses, which resulted in failure of the transducer filters. An improved housing design for miniature pore pressure transducers is required.

10.3.1.2 The Pile Hammer

The methods developed in this study to design a laboratory pile hammer were successful. The electro-pneumatic control system provided excellent control of the hammer. The hammer delivered a repeatable rated energy of 156 ft-lb with a 57% efficiency. The maximum impact velocity and operating frequency were 8.2 ft/sec and 1.2 Hz, respectively. Energy was delivered in a manner consistent with pile driving in the field.

Most of the energy losses in the hammer occurred due to the development of a cushioning back pressure ahead of the ram. Future hammer designs should attempt to increase hammer efficiency by mounting the control valve on the hammer, while isolating it from shock and vibration. Additionally, air blow-by was found to be highly beneficial, and should be allowed as long as sufficient air supply is available.

10.3.1.3 The Pressure Chamber System

An experimental facility for performing foundations research under controlled laboratory conditions has been developed. The facility has been used, successfully, in research on pipe piles and suction piles. The facility is designed to facilitate research on other types of foundations, such as shallow foundations.

10.3.1.4 Modeling Capabilities of the Developed Apparatus

The experimental apparatus developed in this study should be considered primarily as a small prototype which is used to study pile behavior. The developed apparatus satisfies most, but not all, the requirements of similitude for model test requirements. The primary inadequacy of the apparatus is its inability to model the gradual increase in soil stresses with depth, which limits the modeling capabilities to modeling a pile-soil element at a given depth. The pressure chamber can model the effective stresses in soil strata up to 120 ft. below the mud line. The hammer can model both dynamic and diffusion times of typical piles.

10.3.2 Conclusions Regarding the Behavior of Piles in Sand

The results of the testing program provide a number of important insights into the behavior of pipe piles during installation and loading, as follows:

10.3.2.1 Pile Plugging

Piles plugged during load tests and during installation by pushing, but not during driving. Formation of the plug during installation by pushing did not significantly change the penetration resistance. Furthermore, although the plug formed suddenly there was no sudden change in penetration resistance. When a plug formed, 80% of the end bearing was transferred to the internal surface of the pile in a length equal (or less than) one pile diameter from the tip. The remaining 20% of the end bearing was transferred to the pile in the second pile diameter.

Plugging results in the formation of very large radial stresses inside the pile. These stresses were larger for pushed piles than for driven piles. The radial stresses were also larger for piles in dry sand than saturated sand.

During pull-out, very small pile movements were required to release the plug stresses. The plug was retained in piles installed in saturated sand but it slipped in piles installed in dry sand.

10.3.2.2 Buildup of Pore Pressures during Installation and Loading

Pile driving resulted in a pore-pressure surge near the tip. The rise and decay times of the pressure surge were each approximately 0.1 sec long. The pressure surge resulted in a small build up of ambient pore pressure during driving. The measured pore pressures were approximately 7.5% of the initial in situ effective stresses. Evidence also exists that the pore-pressure inside the plug was higher than the hydrostatic value.

A small build up of pore-pressure was measured during installation by pushing and during load tests. The measured pore pressures ranged between 4% and 10% of the initial in situ effective stresses. Excess pore pressures dissipated within 20 minutes from installation.

10.3.2.3 Load Transfer

For piles pushed in dry sand, skin friction was greatly influenced by residual stresses. The effects of residual stresses were prominent until the load used to install the pile by pushing was exceeded. Residual stresses appear to have lesser effects on piles installed in saturated sand.

Load transfer outside the pile appeared to increase with depth despite the fact that the pressure chamber had initial isotropic stresses.

10.4 Recommendations for Future Research

The experimental results of this research contribute to a better understanding of pile behavior, but they also raise additional questions. Future experimental research on pile foundations should continue to use the techniques developed in this study, particularly fast electro-pneumatic hammers and the double-wall piles.

The author sincerely hopes that the apparatus designs presented in this monograph would be beneficial to foundation engineering researchers. In the short term, the presented designs should be used by researchers to identify the physical mechanisms which control the behavior of piles during installation and loading. In the long term the measured phenomena should provide a framework for developing a theoretically sound model to predict the capacity of piles.

References

American Petroleum Institute (API): Recommended Practice for Planning, Designing, and Constructing Fixed Offshore Platforms, Working Stress Design, 21'st edition (RP 2A-WSD), American Petroleum Institute (2000)

References

1. Al-Douri, R.H., Hull, T.S., Poulos, H.G.: Influence of Test Chamber Boundary Conditions on Sand Bed Response. Geotechnical Testing Journal, GTJODJ 16(4), 550–562 (1993)
2. Al-Shafei, K.A., Cox, W.R., Helfrich, S.C.: Pile Load Tests in Dense Sand - Analysis of Static Test Results. In: Proc. Offshore Technology Conf., pp. 83–103 (1994)
3. Alansari, O.: Capacity and Behavior of Steep Pipe Piles in Dry Sand, Dissertation, The University of Texas at Austin (1999)
4. Allard, M.A.: Soil Stress Field Around driven Piles, Dissertation, California Institute of Technology (1990)
5. American Institute of Steel Construction, Manual of Steel Construction-Load and Resistance Factor Design, 1st edn. (1986)
6. American Petroleum Institute (API): Recommended Practice for Planning, Designing, and Constructing Fixed Offshore Platforms, Working Stress Design, 21st edn. (RP 2A-WSD), American Petroleum Institute (2000)
7. Anderson, D.G., Stokoe, K.H.: Shear Modulus a Time Dependent Soil Property. Dynamic Geotechnical Testing, ASTM, 66–90 (1978)
8. Armishaw, J.W., Cox, D.W.: The Effects of Changes in Pore Pressures on The Carrying Capacities and Settlements of Driven Piles End Bearing in a Sand and Gravel Stratum. In: Recent Developments in The Design and Construction of Piles, pp. 227–236. Institute of Civil Engineers, London (1979)
9. Aschenbrenner, T.B., Olson, R.E.: Prediction of Settlement of Single Piles in Clay. In: Analysis and Design of Pile Foundations, pp. 41–58. ASCE Special Technical Publication (1984)
10. ASTM, Annual Book of Standards, Vol. 4.08, ASTM 1916 Race St., Philadelphia, PA 19103 (1993)
11. Beckwith, T.G., Marangoni, R.D.: Mechanical Measurements, 4th edn. Addison-Wesley Publishing Company (1990)
12. Bernardes, G.D.: Dynamic and Static Testing of Large Model Piles in Sand, Ph.D. Thesis, Norwegian Institute of Technology (NTH), University of Trondheim (1989)
13. Bishop, A.W., Henkel, D.J.: The Measurements of Soil Properties in the Triaxial Test, 2nd edn. Edward Arnold, London (1969)
14. Blake, A.: Practical Stress Analysis in Engineering Design, 2nd edn. Marcel Dekker, Inc., NY (1990)
15. Bolton, M.D., Lau, C.K.: Scale Effects Arising from Particle Size. In: Proc. Centrifuge 1988, Balkema, Rotterdam (1988)
16. Bond, A.J., Jardine, R.J., Dalton, J.C.P.: Design and Performance of the Imperial College Instrumented Pile. Geotechnical Testing Journal, ASTM 14(4), 413–424 (1991)
17. Briaud, J., Audibert, J.: Some Thoughts on API RP2A for Vertically Loaded Piles. In: Proc. 22nd Offshore Technology Conf., pp. 9–16 (1990), doi:10.4043/6418-MS
18. Briaud, J.L., Anderson, J.: Evaluation of API Method Using 98 vertical pile load tests. In: Proc. 19th Offshore Technology Conf. (1987)

19. Brucy, F., Meunier, J., Nauroy, J.F.: Behavior of Pile Plug in Sandy Soils During and After Driving. In: Proc. Offshore Technology Conf., Paper No: 6514-MS (1991), doi:10.4043/6514-MS
20. Butler, J., Caliendo, J., Goble, G.: Comparison of SPT Energy Measurement Methods. In: Robertson, P., Mayne, P. (eds.) Proc., Geotechnical Site Characterization, Balkema, pp. 901–905 (1998)
21. Chellis, R.D.: Pile Foundations, 2nd edn. McGraw Hill Book Company (1961)
22. Chen, D.D.S., Toto, J.V., Wong, I.H.: Field Evaluation of Hammer Efficiency and Pile Driving Criteria. In: Fuller, F.M. (ed.) Proc. ASCE Piling Symposium, Atlanta, GA (1979)
23. Cheney, J.A., Fragaszy, R.J.: The Centrifuge as a Research Tool. Geotechnical Testing Journal, GTJODJ 7(4), 182–187 (1984)
24. Choi, Y., O'Neil, M.: Soil Plugging and Relaxation in Pipe Pile During Earthquake Motion. ASCE J. Geotechnical and Geoenvironmental Engr. 123(10), 975–982 (1997)
25. Chow, F., Jardine, R., Brucy, F., Nauroy, J.: Effect of Time on Capacity of Pipe Piles in Dense Marine Sand. ASCE J. Geotechnical and Geoenvironmental Engr. 124(3), 254–264 (1998)
26. Cole, C.T., Sundman, J.: UNIX in Real Time, What it Takes to Make the Grade, UNIX Review (1985)
27. Considine, D.M.: Process Instruments and Controls Handbook. McGraw-Hill Book Company (1957)
28. Coop, R., Wroth, C.: Field Studies of an Instrumented Model Pile in Clay. Géotechnique 39(4), 679–696 (1989)
29. Coyle, H.M., Castelo, R.R.: New Design Correlations for Piles in Sand. Journal of Soil Mechanics and Foundation Division, ASCE 107, 956–986 (1981)
30. Coyle, H.M., Sulaiman, I.H.: Skin Friction for Steel Piles in Sand. J. Soil Mech. and Found. Div., ASCE 93(SM6), 261–278 (1967)
31. Craig, W.H.: Modeling of Pile Installation in Centrifuge Experiments. In: Proc. 11th Int. Conf. on Soil Mechanics and Foundation Engineerting, vol. 2, pp. 1101–1104 (1985)
32. D'Appolonia, D.J., Lambe, T.W.: Performance of Four Foundations on End Bearing Piles. Journal of Geotechnical Engineering 8, 434–455 (1971)
33. Darrag, A.A.: Capacity of Driven Piles in Cohesionless Soils Including Residual Stresses, Ph.D. Dissertation, School of Civil Engineering, Purdue University (1987)
34. Datta, M.: Pore Pressure Development During Pile Driving and Its Influence on Driving Resistance. In: Proc. Geotechnical Practice In Offshore Engineering, Austin, Texas. ASCE (1983)
35. Davisson, M.T.: High Capacity Piles. In: Soil Mech. Div., Ill. Sect., pp. 81–112. ASCE, Chicago (1973)
36. de Beer, E.E.: Bearing Capacity and Settlement of Shallow Foundations on Sand. In: Proc. Bearing Capacity and Settlement of Foundations, Department of Civil Engineering, Duke University, Durham, N.C., pp. 15–33 (1967)
37. De Nicola, A., Randolph, M.: The Plugging Behaviour of Driven and Jacked Piles in Sand. Geotechnique 47(4), 841–856 (1997)
38. Deeks, A.J., Randolph, M.F.: Analytical Modeling of Hammer Impact for Pile Driving. Int. Journal for Numerical and Analytical Methods in Geomechanics 17, 279–302 (1993)

39. Dennis Jr., N.D., Olson, R.E.: Axial Capacity of Steel Pipe Piles in Clay. In: Proc. Geotechnical Practice in Offshore Engineering, ASCE, Geotechnical Engr. Div. Specialty Conf., University of Texas at Austin, pp. 370–388 (1983a)
40. Dennis Jr., N.D., Olson, R.E.: Axial Capacity of Steel Pipe Piles in Sand. In: Proc. Geotechnical Practice in Offshore Engineering., ASCE Geotechnical Engr. Div. Specialty Conf., Univ. of Texas at Austin, pp. 389–402 (1983b)
41. Dow Corning, Information About High Technology Silicone Materials, Dow Corning 3145 rtv Adhesive Sealant, Dow Corning Corporation, Midland, MI (1990)
42. Dow Corning, Information About High Technology Silicone Materials, SYLGARD 567 Primerless Silicone Encapsulant, Dow Corning Corporation, Midland, MI (1993)
43. Dow Corning, Materials for High Technology Applications, Dow Corning Corporation, Midland, MI (1991)
44. Edwards, T.C.: Piling Analysis Wave Equation Computer Program Utilization Manual, Texas Transportation Institute, Research Project Report 33-11, Texas A&M University, p. 40 (1967)
45. El-Mabsout, M.: A Finite Element Model for the Analysis of Pile Driving, Ph.D. Dissertation, University of Texas at Austin (1991)
46. Ellenger, W.(Bill): Proprietary Test Data, Parker Hannifin Corporation, Richland, Michigan 49083 (1993)
47. Farrar, J.: Summary of Standard Penetration Test Energy Measurements Experience. In: Robertson, P., Mayne, P. (eds.) Proc., Geotechnical Site Characterization, Balkema, pp. 919–926 (1998)
48. Fleming, W.G.K., Weltman, A.J., Randolph, M.F., Elson, W.K.: Piling Engineering. John Wiley and Sons, NY (1985)
49. Focht, J.A., O'Neill, M.: International State of The Practice for Design and Installation of Axially Loaded Piles. In: Proc. 11th International Conf. on Soil Mechanics and Foundation Engineering, vol. 1, pp. 187–209 (1985)
50. Fouse, J.L.: Group Behavior of Axially-Loaded Piles in Sand, Master Thesis, The University of Texas at Austin (1984)
51. Fuller, F.M.: Engineering of Pile Installations McGraw-Hill Book Company (1983)
52. Furlow, C.R.: Pile Tests, Jonesville Lock and Dam Quachita and Black Rivers Arkansas and Louisiana, In: Technical Report S-68-10, U.S. Army Engineer Waterways Experiment Station (1968)
53. Gavin, K., Lehane, B.: The Shaft Capacity of Pipe Piles in Sand. Canadian Geotechnical J. 40, 36–45 (2002)
54. GE Silicones, Tech Note CDS4518, General Electric Company, Wayford, NY (1990)
55. GE Silicones, Two Component RTV Silicone Rubber Compounds for Industrial Applications, General Electric Company, Wayford, NY (1991)
56. Ghionna, V.N., Jamiolkowski, M.: A Critical Appraisal of Calibration Testing in Sands. In: Proc. of the 1st Int. Symposium on Calibration Chamber Testing, Potsdam, NY, pp. 13–39 (1991)
57. Gills, J.J.: Development and Testing of a Device Capable of Placing Model Piles by Driving and Pushing in the Centrifuge, Ph.D. Dissertation, Univ. of Florida (1988)
58. Go, V.: Axial Static Load Capacity of Untapered Piles in Sand, M.Sc. in Engr. Thesis, The University of Texas at Austin (1990)
59. Goble, G.G., Rausche, F., Likens, G.: The Analysis of Pile Driving. In: A State of the Art, Int. Seminar on The Application of Stress Wave Theory on Piles, Stockholm, Balkema, pp. 131–161 (1980)

60. Goble, G.G., Rausche, F.: Wave Equation Analysis of Pile Driving - WEAP Program, Report, Prepared for the U.S. Department of Transportation, Federal Highway Administration, Implementation Division, Office of Research and Development (1976)
61. Goble, G.G.: A Pile Inspector's Guide to Hammers and Pile Driving Systems, Pennsylvania Department of Transportation (1987)
62. GRL, Documentation of WEAP 1987 Computer Program, Vol. 1-4 (1988)
63. Hajduk, E., Lin, G., Adams, J., Ledford, D.: Experiences with Open Ended Pipe Pile Plugging in The Atlantic Coastal Plain. In: Iskander, M., et al. (eds.) Contemporary Topics in Deep Foundations (GSP 158), pp. 1–10 (2009), doi:10.1061/40902(221)38
64. Helfrich, S.C., Wiltsie, E.A., Cox, W.R., Al-Shafei, K.A.: Pile Load Tests on Dense Sand: Planning, Instrumentation, and Results. In: Proc. 7th Offshore Technology Conference, vol. 1 (1985)
65. Hight, D., Lawrence, D., Farquhar, G., Mulligan, G., Gue, S., Potts, D.: Evidence for scale effects in the end bearing capacity of open-ended piles in sand, Paper No. 7975-MS (1996), doi:10.4043/7975-MS
66. Holubec, I., D'Appolonia, E.: Effect of Particle Shape on the Engineering Properties of Granular Soils, Evaluation of Relative Density and its Role in Geotechnical Projects Involving Cohesionless Soils. In: ASTM STP 523, American Society of Testing and Materials, pp. 304–318 (1973)
67. Hossain, M.K., Briaud, J.L.: Pipe Piles in Sand - An Improvement for the API RP 2A. In: Proc. Offshore Technology Conf. (1992)
68. Huntley, H.E.: Dimensional Analysis. MacDonald & Co (Publishers) LTD., London (1952)
69. Hvorslev, M.J.: Subsurface Exploration and Sampling of Soils for Civil Engineering Purposes, ASCE Research Project of the Committee on Sampling and Foundations, ASCE (1949)
70. Igoe, D., Doherty, P., Gavin, K.: The Development and Testing of an Instrumented Open-Ended Model Pile. ASTM Geotechnical Testing J. 33(1), 72–82 (2010)
71. IHC, IHC Hydro Hammer, PO Box 26, 2960 AA Kinderdijk, The Netherlands (1988)
72. Ishibashi, I.: Effect of Grain Characteristics on Liquefaction Potential – In Search of Standard Sand for Cyclic Strength. Geotechnical Testing Journal, GTJODJ 8(3) (1985)
73. Iskander, M.: An Experimental Facility to Model the Behavior of Steel Pipe Piles in Sand, Dissertation, The University of Texas at Austin (1995)
74. Iskander, M., Olson, R., Bay, J.: Design and Performance of an Electro-Pneumatic Pile Hammer for Laboratory Applications. Geotechnical Testing J., GTJODJ 24(1), 72–82 (2000)
75. Iskander, M.G., Olson, R.E.: Review of API Guidelines for Pipe Piles in Sand. In: Proc. Civil Engineering in the Oceans V, pp. 798–812. ASCE (1992)
76. Jones, W.C.: Axial Capacity of Suction Piles in Sand, M.Sc. Thesis, The University of Texas at Austin (1994)
77. Khan, N.I., Templeton, J.S., O'Neill, M.W.: Skin Friction Distribution on Piles in Sand. In: Proc. Civil Engineering in The Oceans V, pp. 783–797. ASCE (1992)
78. Kishida, H., Isemoto, N.: Behavior of Sand Plugs in Open Ended Steel Pipe Piles. In: Proc. 9th Int. Conf. on Soil Mechanics and Foundation Engineering, Tokyo, pp. 601–604 (1977)

79. Kishida, H., Uesugi, M., Susumu, M.: Behavior of Dry Sands in Steel Pipe Piles. In: Proc. 8th Southeast Asian Geotechnical Conf., Kuala Lumpur (1985)
80. Kishida, H.: The Ultimate Bearing Capacity of Pipe Piles in Sand. In: Proc. Third Asian Regional Conf. of Soil Mechanics and Foundation Engineering, vol. 1, pp. 196–199 (1967)
81. Kraft, L.: Computing Axial Pile Capacity in Sands for Offshore Conditions. Marine Geotechnology, 61–92 (1990)
82. Kraft, L.: Performance of Axially Loaded Pipe Piles in Sand. Journal of Geotechnical Engineering, 272–296 (1991)
83. Kulhawy, F.H.: Limiting Tip and Side Resistance, Fact or Fallacy? In: Meyer, J.R. (ed.) Proc. ASCE Symposium on Analysis and Design of Pile Foundation, San Francisco, CA (1984)
84. Kutter, B.K., Sathialingam, N., Hermann, L.: Effects of Arching on Response Time of Miniature Pore Pressure Transducer in Clay. Geotechnical Testing Journal, astm, gtjodj 13(3), 164–178 (1990)
85. Lacasse, S., Goulois, A.M.: Uncertainty in API Parameters for Prediction of Axial Capacity of Driven Piles in Sand. In: Proc. 21st Offshore Technology Conf., Houston, OTC paper 6001, pp. 353–358 (1989)
86. Lacasse, S.: Uncertainties in Offshore Geotechnical Engineering. Int. Survey of API-RP 2A Design Parameters for Axial Capacity of Driven Piles in Sand, Norwegian Geotechnical Institute, Report No: 85307-14 (1998)
87. Lacy, H.S.: Pile Integrity and Capacity Determined by Redriving. In: Proc. 10th Int. Conf. on Soil Mechanics and Foundation Engineering, Stockholm, vol. 2, pp. 781–786 (1981)
88. Langhaar, H.L.: Dimensional Analysis and Theory of Models, 4th edn. John Wiley & Sons, NY (1960)
89. Lee, C.Y., Poulos, H.G.: Influence of Excess Pore Pressures on Axial Offshore Pile Response, Research Report No: R577, The University of Sydney (1988)
90. Lee, F.: Frequency Response of Diaphragm Pore Pressure Transducers in Dynamic Centrifuge Model Tests. Geotechnical Testing Journal, gtjodj 13(3), 201–207 (1990)
91. Lehane, B., Gavin, K.: Base Resistance of Jacked Pipe Piles in Sand. ASCE J. Geotechnical and Geoenvironmental Engineering 127(6), 473–480 (2001)
92. Lehane, B., Randolph, M.: Evaluation of a Minimum Base Resistance for Driven Pipe Piles in Siliceous Sand. ASCE J. Geotechnical and Geoenvironmental Engineering 128(3), 198–205 (2002)
93. Lehane, B.M., Jardine, R.J., Bond, A.J., Frank, R.: Mechanism of Shaft Friction in Sand From Instrumented Pile Tests. Journal of Geotechnical Engineering, ASCE 119, 19–35 (1993)
94. Lehane, B.M., Jardine, R.J.: Shaft Capacity of Driven Piles in Sand: A New Design Approach. In: Proc. 7th Int. Conf. on the Behavior of Offshore Structures (BOSS 1994), vol. 1, pp. 23–35 (1994)
95. Leong, E., Randolp, M.: h Finite Element Analysis of Soil Plug Response. Int. Journal of Numerical and Analytical Methods in Geomechanics 15, 121–141 (1991)
96. Lings, M.L.: The Skin Friction of Driven Piles in Sand, M.Sc. Thesis, Imperial College of Science and Technology, University of London (1985)
97. Lo, K.Y., Stermac, A.G.: Induced Pore Pressures During Pile Driving Operations. In: Proc. 6th Int. Conf. on Soil Mechanics and Foundation Engineering, vol. 2, pp. 285–289 (1965)

98. Lowery, L.L., Edwards, L.C., Hirsch, T.J.: Use of The Wave Equation to Predict Soil Resistance on a Pile During Driving, Texas Transportation Institute, Research Report 33-10 (1968)
99. Malahotra, S.: Effect of Wall Thickness on Plugging of Open Ended Steel Pipe Piles in Sand. In: Contemporary Issues in Deep Foundations, GSP 158 (2007), doi:10.1061/40902(221)36
100. Malhotra, S.: Development of a Model Pile-Driving Facility, M.Sc. Thesis, The University of Texas at Austin (1991)
101. Masscomp, MC 5400 Hardware Maintenance Manual, Concurrent Computer Corporation, 2 Crescent Place, Ocean Port, NJ 07757 (1985)
102. McClelland, B., Focht, J.: Padre Island Load Tests, Project Report, McClelland Engineers (Now Fugro-McClelland Marine Geosciences), 6100 Hilcroft, Houston, Texas (1955)
103. Measurements Group Inc., Bulletin B-129-7, Surface Preparation for Strain Gage Bonding, Micro-Measurements Division, P.O. Box 27777, Raleigh, NC 27611 (1976)
104. Measurements Group Inc., Bulletin B-137-13, Strain Gage Applications with M-Bond AE-10/15 and M-Bond GA-2 Adhesive Systems, Micro-Measurements Division, P.O. Box 27777, Raleigh, NC 27611 (1979a)
105. Measurements Group Inc., Catalog 500, Part A, Strain Gage Listings, Micro-Measurements Division, P.O. Box 27777, Raleigh, NC 27611 (1988a)
106. Measurements Group Inc., Catalog 500, Part B, Strain Gage Technical Data, Micro-Measurements Division, P.O. Box 27777, Raleigh, NC 27611 (1988b)
107. Measurements Group Inc., Tech Note TN-502, Optimizing Strain Gage Excitation Levels, Micro-Measurements Division, P.O. Box 27777, Raleigh, NC 27611 (1979b)
108. Measurements Group Inc., Tech Note TT601, Techniques for Bonding Leadwires to Surfaces Experiencing High Centrifugal Forces, Micro-Measurements Division, P.O. Box 27777, Raleigh, NC 27611 (1982)
109. Measurements Group Inc., Tech Note TT607, Strain Gage Installation and Protection in Field Environments, Micro-Measurements Division, P.O. Box 27777, Raleigh, NC 27611 (1983)
110. Measurements Group Inc., Tech Note TT608, Techniques for Attaching Leadwires to Unbonded Strain Gages, Micro-Measurements Division, P.O. Box 27777, Raleigh, NC 27611 (1984)
111. Measurements Group Inc., Tech Note TT609, Strain Gage Soldering Techniques, Micro-Measurements Division, P.O. Box 27777, Raleigh, NC 27611 (1986a)
112. Measurements Group Inc., Tech Note TT610, Strain Gage Clamping Techniques, Micro-Measurements Division, P.O. Box 27777, Raleigh, NC 27611 (1986b)
113. Mehle, J.S.: Centrifuge Modeling of Pile Driving, Master Thesis, Department of Civil, Environmental, and Architectural Engineering, University of Colorado, Boulder Colorado (1989)
114. Menck GMBH, MHU, Offshore Hydraulic Pile Driving Hammers, Werner-Von-Siemens-Strasse 2, D2086 Ellerau, Germany (1990a)
115. Menck GMBH, MRBS, Offshore Steam/Air Pile Driving Hammer, Werner-Von-Siemens-Strasse 2, D2086 Ellerau, Germany (1990b)
116. Mesri, G., Feng, T.W., Benak, J.M.: Post Densification Penetration Resistance of Clean Sands. Journal of Geotechnical Engineering, ASCE (1990)

117. Meyerhof, G.G.: Bearing Capacity and Settlement of Pile Foundations. Journal, Geotechnical Engr. Div., ASCE 102(GT3), 196–228 (1976)
118. Miller, D.S.: Compressible Internal Flow. British Hydromechanics Research Association (BHRA) Fluid Engineering Series, BHRA, vol. 10 (1984)
119. Moe, D., Arveson, H., Holm, O.S.: Friction Bearing Pipe Piles at Calabar Port. In: Proc. 10th Int. Conf. on Soil Mechanics and Foundation Engineering, Stockholm (1981)
120. Moller, B., Bergdahl, U.: Dynamic Pore Pressures During Pile Driving in Fine Sand. In: Proc. Int. Conf. on Soil Mechanics and Foundation Engineering, vol. 2, pp. 791–794 (1981)
121. Murff, J., Raines, R., Randolph, M.: Soil Plug Behavior of Piles in Sand. In: Proc. 22nd Offshore Technology Conf., Houston, TX, pp. 25–32 (1990)
122. Murff, D.: Personal Communications, Exxon Production Research Company, Houston, Texas (1992)
123. Murff, J.D.: Axial Pile Capacity in Sand, Draft Proposal for API 1990 Research Project, Exxon Production Research Company, P.O. Box 2189, Houston, Texas 77252 (1998)
124. Murff, J.D.: Marine Foundation Engineering Class Notes, Texas A&M University (1990)
125. Notingham, L.E.: Use of Quasi-Static Friction Cone Penetrometer Data to Predict Load Capacity of Displacement Piles, Ph.D. Dissertation, University of Florida (1975)
126. Nunez, I.L., Hoadley, P.J., Randolph, M.F., Hulett, J.M.: Driving and Tension Loading of Piles in Sand on a Centrifuge. In: Proc. Centrifuge 1988, Balkema, pp. 353–362 (1988)
127. Nystrom, G.: Finite-Strain Axial Analysis of Piles in Clay. In: Meyer, J.R. (ed.) Proc. ASCE Symposium on Analysis and Design of Pile Foundation, San Francisco, CA (1984)
128. O'Neill, M.: Houston Calibration Chamber: Case Histories. In: Proc. 1st Int. Symposium on Calibration Chamber Testing, Potsdam, NY, pp. 277–288. Elsevier, NY (1991)
129. O'Neill, M.W., Raines, R.D.: Load Transfer for Pipe Piles in Highly Pressured Dense Sand. Journal of Geotechnical Engineering, 1208–1226 (1991)
130. Ogilvie, A.: Measurement of Engineering Properties for Sand Used in Tank Simulation of Offshore Pile Driving, Project Report, NSF Careers for Minority Scholars, The University of Texas at Austin (1995)
131. Olson, R., Iskander, M.: Axial Load Capacity of Un-Tapered Piles In Cohesionless Soils. In: Contemporary Topics in Deep Foundations. GSP No. 185, pp. 231–238. ASCE (2009)
132. Olson, R.E., Al-Shafei, K.S.: Axial Load Capacities of Steel Pipe Piles in Sand. In: Proc. 2nd Intern. Conf. on Case Histories in Geot. Engr., Univ. of Missouri, Rolla, vol. 3, pp. 1731–1738 (1988)
133. Olson, R.E., Dennis, N.D.: Review and Compilation of Pile Load Test Results, Axial Pile Capacity, Final Report to The American Petroleum Institute on Project PRAC 81-29, Geotechnical Engineering Report CR83-4. Department of Civil Engineering, The University of Texas at Austin (1982)

134. Olson, R.E., Iskander, M.G.: Axial Load Capacity of Pipe Piles in Sands. In: Proc. 1st Regional Conf. and Exhibition on Advanced Technology in Civil Engineering, Manama, Bahrain, pp. 383–394. ASCE–SAS (1994)
135. Olson, R.E.: Axial Load Capacity of Steel Pipe Piles. In: Proc. 22nd Offshore Technology Conf., Paper 6419 (1990)
136. Paik, K., Salgado, R.: Determination of Bearing Capacity of Open-Ended Piles in Sand. ASCE J. Geotechnical & Geoenvironmental Engr. 129(1), 46–57 (2003)
137. Paik, K., Salgado, R., Lee, J., Kim, B.: Behavior of Open- and Closed-Ended Piles Driven Into Sands. ASCE J. Geotechnical & Geoenvironmental Engr. 129(4), 296–306 (2003)
138. Paik, K., Salgado, R., Lee, J., Kim, B.: Behavior of Open- and Closed-Ended Piles Driven Into Sands. ASCE J. Geotechnical & Geoenvironmental Engr. 129(4), 296–306 (2003)
139. Paik, K., Salgado, R.: Determination of Bearing Capacity of Open-Ended Piles in Sand. ASCE J. Geotechnical & Geoenvironmental Engr. 129(1), 46–57 (2003)
140. Paik, K.H., Lee, D.R.: Behavior of Soil Plugs in Open Ended Model Piles Driven Into Sands. Marine Geotechnology 11, 353–373 (1993)
141. Paikowsky, S., Whitman, R., Baligh, M.: A New Look at The Phenomena of Offshore Pile Plugging. Marine Geotechnology 8, 213–230 (1989)
142. Paikowsky, S.G.: A Static Evaluation of Soil Plug Behavior with Application To The Pile Plugging Problem, D.Sc. Thesis, MIT (1989)
143. Paikowsky, S.G., Whitman, R.V.: The Effect of Plugging on Pile Performance and Design. Canadian Geotechnical Journal, 429–440 (1990)
144. Paikowsky, S.G.: The Mechanism of Plugging in Sand. In: Proc. 22nd Offshore Technology Conf., pp. 593–604 (1990)
145. Paikowsky, S.G.: A Static Evaluation of Soil Plug Behavior with Application to the Pile Plugging Problem, Ph.D. Thesis, Massachusetts Institute of Technology (1989)
146. Parker, Parker Pneumatic Products Catalog 0107-1, Pneumatic Division, Parker Hannifin Corporation, Richland, Michigan 49083 (1991)
147. Parkin, A.K.: The Calibration of Cone Penetrometers. In: deRuiter, J. (ed.) Proc. Penetration Testing 1988, Balkema, Rotterdam, vol. 1, pp. 221–243 (1988)
148. Parkin, A.K., Lunne, T.: Boundary Effects in the Laboratory Calibration of a Cone Penetrometer in Sand. In: Proc. 2nd European Symposium on Penetration Testing, vol. 2, pp. 761–768 (1982)
149. Pavlicek, R.W.: Axial Tensile Load Capacity of Suction Piles, M.Sc. Thesis, The University of Texas at Austin (1993)
150. Peck, R.B., Hansen, W.E., Thornburn, T.H.: Foundation Engineering, p. 114. John Wiley & Sons, NYC (1974)
151. Pelletier, J.H., Murff, J.D., Young, A.C.: Historical Development and Assessment of the Current API Design Methods for Axially Loaded Pipes. In: Proc. Offshore Technology Conf., Houston, TX, OTC paper 7157, pp. 253–282 (1993)
152. Perry, C.C., Lissner, H.R.: The Strain Gage Primer, 2nd edn. McGraw Hill Book Company (1962)
153. Pileco-Delmag, Corporate Brochures and Hammer Data Sheets, PO Box 16099, Houston, TX 77222 (1992)
154. Platema, G.: The occurrence of Hydrodynamic Stresses in The Pore Water of Sand Layer During Driving of Piles. In: Proc. 2nd Int. Conf. on Soil Mechanics and Foundation Engineering, Rotterdam, vol. 4, pp. 127–128 (1948)

155. Poorooshasb, F.: On Centrifuge Use for Ocean Research. Marine Geotechnology 9, 141–158 (1990)
156. Poulos, H.G., Davis, E.H.: Pile Foundation Analysis and Design. John Wiley and Sons, NY (1968)
157. Rad, N.S., Tumay, M.T.: Factors Affecting Sand Specimen Preparation by Raining. Geotechnical Testing Journal, ASTM 10(1), 31–37 (1987)
158. Rains, R.D., Ugaz, O., O'Neill, M.: Driving Characteristics of Open Towe Piles in Sand. Journal of Geotechnical Engineering, 72–88 (1992)
159. Randolph, M., Leong, E., Hyden, A., Murff, J.: Soil Plug Response in Open Ended Pipe Piles. ASCE J. of Geotechnical & Geoenvironmental Engr., 743–759 (1990)
160. Randolph, M.: Modeling of The Soil Plug Response During Pile Driving. In: Proc. 9th Asian Geotechnical Conf., Bangkok, Thailand, pp. 6-1–6-14 (1987)
161. Randolph, M.F., Leong, E.C., Houlsby, G.T.: One Dimensional Analysis of Soil Plugs in Pipe Piles. Geotechnique 41, 587–598 (1991)
162. Randolph, M.F.: Capacities of Piles Driven into Dense Sands. Presentation to the 11th Conf. of Soil Mechanics and Foundation Engineering, San Francisco, Session 4C (1985)
163. Randolph, M.F.: Capacity of Piles Driven Into Dense Sand. Presentation to the 11th Int. Conf. of Soil Mechanics and Foundation Engineering, San Francisco (1985)
164. Reese, L.C., Seed, H.B.: Pressure Distribution Along Friction Piles. In: Proc. ASTM, vol. 55 (1955)
165. Reese, L.C.: Interaction of Deep Foundations with the Supporting Soils. In: Proc. 1st Indian Conf. in Ocean Engineering, Madrass, India, vol. 2, pp. 143–154 (1981)
166. Robertson, P.K., Woeller, D.J., Gillespie, D.: Evaluation of Excess Pore Pressures and Drainage Conditions Around Driven Piles Using The Cone Penetration Test With Pore Pressure Measurements. Canadian Geotechnical Journal, 249–254 (1990)
167. Robinsky, E.I., Morrison, C.F.: Effect of Shape and Volume on The Capacity of Model Piles in Sand. Canadian Geotechnical Journal 1(4), 189–204 (1964)
168. Robinsky, E.I., Morrison, C.F.: Sand Displacement and Compaction Around Model Friction Piles. Canadian Geotechnical Journal 1(2), 81–93 (1964)
169. Robinsky, E.I., Sagar, W.L., Morrison, C.F.: Effect of Shape and Volume on the Capacity of Model piles in Sand. Canadian Geotechnical Journal 1(4), 189–204 (1964)
170. Schmertmann, J.H.: The Mechanical Aging of Soils, 25th Terzaghi Lecture. ASCE (1989)
171. Schnaid, F., Houlsby, G.T.: An Assessment of Chamber Size Effects in the Calibration of In Situ Tests in Sand. Geotechnique 41(3), 437–445 (1991)
172. Schofield, A.N.: Cambridge Geotechnical Centrifuge, 20th Rankine Lecture. Geotechnique 30(3), 227–268 (1980)
173. Seed, H.B., Lee, K.L.: Liquefaction of Saturated Sands During Cyclic Loading. Journal of the Soil Mechanics and Foundations Division, ASCE 92, 105–134 (1966)
174. Seed, H.B., Tokimatsu, K., Harder, L.F., Chung, R.M.: The Influence of SPT Procedures in Soil Liquefaction Resistance Evaluations, Earthquake Engineering Research Center Report No: UCB/EERC-84/15, College of Engineering, University of California, Berkeley (1984)
175. Sellgren, E.: Friction Piles in Non-Cohesive Soils. Evaluation from Pressuremeter Tests, Ph.D. Dissertation, Chalmers University of Technology, Goteborg, Sweden (1982)

176. Skempton, A.W.: Standard Penetration Test Procedures and the Effects in Sand of Overburn Pressure, Relative Density, Particle Size, Aging, Over consolidation. Geotechnique (3) (1986)
177. Smith, I., Chow, Y.: Three Dimensional Analysis of Pile Drivability. In: Proc. 2nd Int. Conf. on Numerical Methods in Offshore Piling, Austin, Texas, pp. 1–19 (1982)
178. Smith, I., To, P., Wilson, S.: Plugging of Pipe Piles. In: Proc. 3rd Int. Conf. on Numerical Methods in Offshore Piling, Nantes, France, pp. 54–73 (1986)
179. Szechy, C.: Tests with Tubular Piles. Acta Technica, Hungarian Academy of Science 24, 181–219 (1958)
180. Szechy, C.: The Effects of Vibration and Driving Upon the Voids in Granular Soil Surrounding a Pile. In: Proc. 5th Int. Conf. of Soil Mechanics and Foundation Engineering, vol. 2, pp. 161–164 (1961)
181. Tang, W.H., Gilbert, R.B.: Case Study of Offshore Pile System Reliability. In: Proc. 25th Offshore Technology Conf., pp. 677–686 (1993)
182. Terzaghi, K.: Theoretical Soil Mechanics. Wiley, NY (1943)
183. Timoshenko, S., Goodier, J.: Theory of Elasticity. McGraw Hill, New York (1970)
184. Timoshenko, S.P., Gere, J.M.: Theory of Elastic Stability, 2nd edn. McGraw Hill Book Company (1961)
185. Tomlinson, M.J.: Some Effects of Pile Driving on Skin Friction, Behavior of Piles. In: Institution of Civil Engrs., London, pp. 107–114, Response to Discussion on pp. 149–152 (1971)
186. Toolan, F.E., Lings, M.L., Mirza, U.A.: An Appraisal of API RP 2A Recommendations for Determining Skin Friction of Piles in Sand. In: Proc. 22'nd Offshore Technology Conf., Houston, TX, Paper No: OTC 6422 (1990)
187. Trautmann, C.H., Kulhawy, F.H., O'Rourke, T.D.: Sand Density Measurements for Laboratory Studies. Geotechnical Testing Journal 8(4), 159–165 (1985)
188. Tsein, S.I.: Shaft Frictional Resistance of Long Pipe Piles Driven into Dense Sands. In: Proc. 18th Offshore Technology Conf., Houston, TX (1986)
189. Uesugi, M., Kishida, H., Tsubakihara, Y.: Behavior of Sand Particles in Sand-Steel Friction. Soils and Foundations 28(1), 107–118 (1988)
190. Ugaz, O.G.: An Experimental and Numerical Study of Impact Driving of Open-Ended Pipe Piles in Dense Saturated Sand, Dissertation, University of Houston (1988)
191. Veismanis, A.: Laboratory Investigations of Electrical Friction Cone Penetrometers in Sand. In: Proc. European Symposium on Penetration Testing, Stockholm, vol. 2, pp. 407–420 (1974)
192. Vesic, A.S.: Expansion of Cavities in Infinite Soil Mass. J. Soil Mech. Found. Div., ASCE 98(SM3), 265–290 (1972)
193. Vesic, A.S.: A Study of Bearing Capacity of Deep Foundations, Final Report on Project B-189, Georgia Tech University, Atlanta (1967)
194. Vesic, A.S.: Load Transfer in Pile-Soil Systems. In: Fang, Dismuke (eds.) Proc., Conf. on Design and Installation of Pile Foundations and Cellular Structures, pp. 47–73. Envo Publishing Co. (1970)
195. Vesic, A.: Principles of Pile Foundations Design. Soil Mechanics Series, vol. 38. School of Civil Engineering, Duke University (1975)
196. Vines, W.R., Amar, J.: Stress Measurements for Offshore Pile Driving. In: Fuller, F.M. (ed.) Proc. ASCE Piling Symposium, Atlanta, GA (1979)

197. Vulcan Iron Works, Bulletin 65K-1, Specifications for Offshore Pile Hammers, PO Box 5402, Chattanooga, TN 37406 (1993)
198. Warrington, D.C.: Personal Communications, Vice Chairman, Vulcan Iron Works, Inc., PO Box 5402, Chattanooga, TN (1994)
199. Warrington, D.C.: A Proposal for a Simplified Model for the Determination of Dynamic Loads and Stresses During Pile Driving. In: Proc. 19th Offshore Technology Conf., OTC paper No: 5395, Houston, Texas (1987)
200. Wong, D.O.: Design and Analysis of an Apparatus to Simulate Density Stresses in Deep Deposits of Granular Soils, Ph.D. Thesis, University of Houston (1985)
201. York, D., Brusey, W., Clemente, F., Law, S.: Setup and Relaxation in Glacial Sand. ASCE J. Geotechnical Engr. 120(9), 1496–1513 (1994)
202. Young, W.C.: Roark's Formulas for Stress & Strain, 2nd edn. McGraw Hill Book Company (1989)
203. Zappe, R.W.: Valve Selection Handbook, 2nd edn. Gulf Publishing Company, P.O. Box 2608, Houston, Texas (1987)
204. Zelikson, A.: Geotechnical Models Using The Hydraulic Gradient Similarity Method. Geotechnique 19(4), 495–508 (1969)

Index